"This is a modern-day 'Erin Brockovich' story of Brianne Dressen fighting Big Pharma and Big Government to keep our health, freedoms, and way of life. No other book in this space provides the unique narrative that delivers a critical and honest 'behind-the-scenes' reality. We all need to read this book and wake up to the hidden oppressions and overreach we continue to experience as a result of Covid-era policy."

—Dr. Ken Ruettgers, professor of sociology, former Green Bay Packer, and author

"Someday, I trust, the roll-out of Covid vaccines will be fully exposed as the greatest medical scandal of all time; how devastating side effects including death were an inconvenience to the billion-dollar pharmaceutical industry and political expedience, and therefore, dismissed and suppressed by regulators, medical professionals, and the media alike. This remarkable book makes it personal—an emotional account of radically changed lives, intense suffering, the sense of abandonment, and the fight for recognition and help. It will make you angry, it will bring tears to your eyes but also warm your heart at the resilience, love, and sheer determination that come from shared adversity. Caroline Pover strikes the perfect balance with sensitivity and integrity."

—Mark Sharman, producer, *Safe and Effective: A Second Opinion*

Worth a Shot? is an eye-opening account that goes beyond the statistics, focusing on the heartfelt story of pain, resilience, and survival experienced by those of us suffering post-vaccine adverse reactions. At its core, is book is a call for more love, compassion, and empathy in the face human suffering. It highlights how the world must evolve, fostering ater understanding and respect for individuals who feel unseen or dissed. In a world divided by opinions, this book urges unity through passion, and a movement towards healing, driven by love for one ther. A must read for everyone."

—Brian Howard, UKCVFamily trustee

"This book chronicles Brianne's shocking journey after her efforts to remedy the injuries she sustained as a volunteer in a Covid vaccine trial became an embarrassment for the entire medical-industrial complex. When the profit and power interests of Big Government and Big Pharma collide with patient health, the drug injury is just the first step in an ago-nizing ordeal. Every American needs to know Bri's account of what hap pens *after* the injury."

—**Robert F. Kennedy,**

"Many were misled by corporations, government agencies, and me professionals during the pandemic. From clinical trials to injury com sation, every American should know how the government is failing who have been harmed by pharmaceuticals."

—**Congressman Thomas**

"According to disinformation put out by the medical establishmen fake 'fact checkers,' Big Tech giants, and some top political figures Dressen simply doesn't exist. *Worth a Shot?* will prove once and f she not only exists but is making sure her gripping story is heard

—**Sharyl Attkisson, bestselling author of *Follow***

"It would take incredible willpower and strength to simply s you live with the sensation of being electrocuted every second and yet Bri has found the courage to not only survive but And she has fought tirelessly, not only to help herself but t injured individuals across the globe. This book is just one myriad ways Bri has fought and continues to fight for wha tant: truth and justice. May it serve as a lesson to us all."

—**Aaron Siri, civil rights and**

"This book is a wake-up call about the gross abuse of 'healthcare.' Sadly, there will be encyclopedias filled soon. Brianne is bravely leading the charge so we wo

—**Drea de Matteo, actress and cofoun**

Worth a Shot?

Worth a Shot?

SECRETS OF THE CLINICAL TRIAL PARTICIPANT
WHO INSPIRED A GLOBAL MOVEMENT—
BRIANNE DRESSEN'S STORY

CAROLINE POVER

Foreword by Brianne Dressen

Skyhorse Publishing

Skyhorse Publishing books may be purchased in bulk at special discounts for sales promotion, corporate gifts, fund-raising, or educational purposes. Special editions can also be created to specifications. For details, contact the Special Sales Department, Skyhorse Publishing, 307 West 36th Street, 11th Floor, New York, NY 10018 or info@skyhorsepublishing.com.

Skyhorse® and Skyhorse Publishing® are registered trademarks of Skyhorse Publishing, Inc.®, a Delaware corporation.

Visit our website at www.skyhorsepublishing.com.

Please follow our publisher Tony Lyons on Instagram @tonylyonsisuncertain.

10 9 8 7 6 5 4 3 2

Library of Congress Cataloging-in-Publication Data is available on file.

Hardcover ISBN: 978-1-5107-8346-1
eBook ISBN: 978-1-5107-8348-5

Cover design by Brian Peterson
Cover photo by Melissa Jo Hudson

Printed in the United States of America

*This book is dedicated to
all those injured by the Covid vaccine,
those we have lost because of it,
and all who love them.*

CONTENTS

FOREWORD

BY BRIANNE DRESSEN

Like the majority of the world, I entered 2020 with a sense of trepidation; uncertain of how life in general seemed to be changing and unsure of what the future may hold. For all of us. Little did I know, that not only was my own life about to change forever, but also my entire perspective of the world was about to evolve, thereby exposing the ugliest of humanity . . . but also its true beauty.

The following year, the health I had enjoyed and the world as I knew it faded further and further away from me. My transformation was coming about because of a process *so* painful, that I would never wish it upon anyone. Trying to make sense of my new life as well as this strange new world to which I had previously been blind was a nightmare; to say that I struggled is an understatement. I was so lost.

At the height of my struggles, I found myself on a call with Caroline Pover, a feisty Brit who wanted all the facts and none of the nonsense. She was an accomplished writer, determined to navigate the issue from a position of love. People were reading her posts on social media and gaining comfort from them, regardless of their political views.

I joined the call with her from my bed, frail and underweight, and was pleased to find someone on the other end of the monitor who was so much more like me than I had expected. She too was suffering; she too was quickly learning about this new world. But her bright smile, colorful language, and ridiculous sense of humor helped me realize I was right at home with her—and maybe a little too much. With Caroline, I felt safe enough to open up about *everything*. About all the ugliness. She helped me find the beauty that existed there, helped me face the uncomfortable questions, and helped me to be honest with myself as well as with her. She became a critical part in illuminating this new life for me.

There have been so many narratives that have been shouted from the rooftops of all camps since the beginning of this mess. I have seen them all. People are resolute in what they think the "truth" really is. I observed all of the commentaries, scientific reviews, and press releases, thinking that none of them matched my own perspective on the events of the past four or five years. I kept wondering . . . maybe next year there will be something? But year after year, there was just more noise, more anger, more judgment, more polarization. None of the "approved" publications seemed to resonate with me nor the many others like me. We are truly the silenced of this time.

It is now time to give a voice to the silenced.

My instincts told me that Caroline was the only person in the world who could provide that voice. I asked her to write this book not only after reading her award-winning other work, but also knowing that, as an advocate herself, she understands this perspective better than anyone. She too is living this experience, breathing this same air, and feeling the same pain. She too is celebrating the same wins, however small they may be and however long they may take. While we spent hours and hours on calls working through each chapter that she had thoughtfully planned, it became clear there was a lot that would go into this book that I didn't need to explain to her. She already knew what needed to be said.

It was very evident to me after reading *Worth a Shot?* for the first time that my instincts were right. There truly was nobody better suited for writing this book than Caroline. She understood the importance of getting this book just right, to reflect not just our own personal experiences but to provide a voice to the suffering of the many others like us. She knew how to be that voice with integrity, compassion, and the human perspective that is so often conveniently omitted from this conversation.

Caroline walks you through my journey in a way that hopefully will open your eyes, without you needing to walk those painful steps yourself. Everyone feels like they have "heard it all" and they know all there is to know, but this is the story that has yet to be told … and has yet to be heard. This story is the one that you have not been *allowed* to be told. We hope that this book helps take a fresh look at the information you have access to, and to ask questions about the individuals you don't.

Brianne Dressen
September 2024

AUTHOR ACKNOWLEDGMENTS

My first thanks goes to my nephew, Jack Smith. As soon as I told you about this project you stepped up to dedicate all the extra hours you possibly could in order to keep my other work going. You immediately understood the importance of this book and created the space for me to focus on it. Your presence in my life is such a blessing. Thank you.

Mum, you filled in with everything else during the final three weeks of writing, taking care of the laundry, dishes, doggies, and all that happens in between. I love you, Schmarm!

My fellow UKCVFamily trustees: Charlet Crichton, Brian Howard, Claire Parham, and Sarah Bell—for giving me permission to focus on this book when we have so much to do. Thank you for everything you all do, every single day. I hope I have made you and "the fam" proud. I am so very proud of you.

Heath Rose, my absolute favourite draft reader . . . thank you for taking the time to read yet another manuscript of mine when your own life is so busy. Your comments are always incredibly valuable and very much appreciated.

Hector Carosso and the rest of the Skyhorse Publishing team . . . thank you for all of your efforts to treat this story with the gentle respect that it requires. Hector, I have been humbled by your compliments about my writing. Thank you.

To my healthcare practitioners, especially my beloved "witchy" Christine, not just for taking care of me while I was writing this, but also for taking care of me during the past three years; getting me to the point where I was physically and mentally strong enough to write this book. I will never be able to appropriately thank you for what you have done for me.

And Bri, my dear friend . . . where do I start? Thank you for sharing everything with me. Thank you for trusting me with the minutes, hours, days, weeks, months, and years of your life where you were at your most vulnerable. Thank you for being brave enough to share that vulnerability with the world. The world needs you, and it needs to know your story. What an absolute *honour* to have been asked to write it.

INTRODUCTION

I didn't believe that Brianne Dressen existed.

It was the spring of 2021, and I had been medically diagnosed with an adverse reaction to the Covid vaccine, about which I subsequently started writing on my social media. I was bombarded with messages from people giving me all sorts of unhelpful opinions on what was happening to my body, and feeling the need to tell me that I was going to die soon. I didn't know what was true and who to trust anymore—about anything—which is a very disconcerting way to move forward in life. So I decided two things: firstly, I would only believe and trust what I could see right in front of me or what someone I knew had personally experienced; this became my way of determining the truth. Secondly, instead of engaging in long arguments either for or against the Covid vaccine, I would just tell people to look at the clinical trial data and make up their own minds.

Then somebody commented on one of my posts, telling me that they knew a woman who was dropped from a clinical trial after she had an adverse reaction. They said that the clinical trial data couldn't be trusted. "Introduce me to her," I responded, assuming the commenter was lying. I was tired of fearmongering people making claims they couldn't prove. But it wasn't a lie. They *did* know a woman who'd been dropped from a trial. And I knew I had to speak to her. I needed to look her in the eye. I needed to know the truth.

That woman was Brianne Dressen. And her story shocked me to my core. I never imagined that three years later she would be asking me to write it.

I am no stranger to immersing myself in other people's pain. I spent almost a decade going back and forth between my home in the Cotswolds, UK, and a remote village in the Northeast of Japan, living within a community that had been devastated by the 2011 tsunami. I dedicated my time to fundraising for them, managing projects to help the community rebuild, listening to whatever they wanted to say about the disaster, and absorbing the energy of the place and its people. It was a life-changing experience that I ended up documenting in an award-winning memoir. The people of Oshika uncovered an ability in me to write about people's pain that I didn't know I had.

When I started writing publicly about the pain experienced by the vaccine-injured, I was surprised by the response it generated, from both injured and non-injured alike. People were surprised by my writing too. Surprised, and sometimes annoyed, because they couldn't fit me into a category—wondering how could I be so reasonable about it all? Was I for or against it? Was I angry or at peace? Was I blaming others or myself? When in fact, I wasn't thinking much about *any* of those things. Most of the time I was just thinking about how to survive my own pain. But I was also thinking about how to move forward *despite* the pain, in a way that created more compassionate conversations and communities. I was thinking about the world I wanted to live in and what things I could do to help create that world.

This book is one of those things.

Worth a Shot? is based on extensive conversations I have personally had with Bri, during which she opened up about parts of her life that have not been covered in the numerous interviews she has given during the past few years. I also researched many mainstream news articles, papers published in scientific journals, and US government reports. I have seen pages and pages of emails between federal health agency representatives and Bri as well as other American vaccine-injured. I have read multiple materials produced by AstraZeneca themselves, or people involved in the development of its Covid vaccine.

This book is based on real events that happened to real people, and it contains many verifiable facts; however, the book is written from Bri's perspective, so the interpretations and emotions involved in reflecting on those events are hers, and hers alone. I acknowledge that some people involved in those events may have different perspectives on and emotions associated with them. The vast majority of names mentioned in this book are names of real people, from whom we sought permission to mention.

The book is brutally honest about a lot of things, including suicide. Chapter 4 may be especially triggering for readers, but I stand firmly by my decision to be so explicit in this chapter. My hope is that this chapter will provide an insight for someone whose loved one is suicidal for any reason (not just related to vaccine injury), and thereby give you perhaps an increased understanding of what your loved one is struggling with. I also hope that this chapter might help bring just a tiny bit of healing to someone who has lost a loved one to suicide. If you, the reader, is suicidal, I hope the knowledge that others have felt the same way as you helps your feelings become less overwhelming. All readers should approach Chapter 4 with caution.

Worth a Shot? should not be treated as a medical book in any way. It does not provide any treatment plans, although it does refer to treatments that have helped Bri and/or other vaccine-injured people. If the reader wants to try any of the protocols mentioned in here, they should first do their own research and seek expert advice and support, but as you will see within these very pages, sometimes the experts are few and far between. Bri and I share the belief that we all need to take responsibility for our own healing, whatever that may mean to each of us.

In the interests of full disclosure, I was paid an advance to write this book, as is standard practice when commissioned. This allowed me to concentrate on the writing on an accelerated schedule, and personally it has helped me attain ongoing treatments for my own injury as well as provide support to other injured. The advance came from a benefactor and the publisher, neither of whom have attempted to exert any control over the content.

Both Brianne and I have waived all rights to any royalties generated by the sales of this book. Instead, all proceeds are being donated equally between two registered charities: React19 in the United States, and UKCVFamily in the United Kingdom. These charities ensure that the Covid vaccine-injured are not forgotten.

Thank you, the reader, for not forgetting about us either.

Caroline Pover
September 2024

PART ONE

THE INJURY

CHAPTER 1

THE TRIAL

Everything in her body was telling her not to do it. Short, gasping little breaths, the pounding in her heart, the tensing of her muscles, and thoughts racing as she faced what she was about to do. She didn't *have* to go ahead with it. She could change her mind. She could turn around and go back at any moment. She could stop this all happening *right now*. But she didn't. She focused on her breathing . . . taking it deeper and deeper . . . she slowed down her heart, cleared the noise in her head, steadied her stance, and calmly found her center before launching herself forward with everything she had; off the fifty-foot cliff and into the freezing waters of Lake Powell below.

Cliff-jumping in Utah was how seven-year-old Brianne Lunt learned to be feisty and fearless, and to push her body far beyond what she thought it was capable of.

It was also how she learned to ignore her own warning signs.

* * *

Brianne, always known as Bri, was the youngest of five kids in a time and a place where kids roamed free. Life was about keeping up with her older brothers and big sister Trish on wherever their adventures would take them. Bri would tag along, knowing that the older kids wouldn't make any concessions for their skinny little sister. *Hurry up, Bri! Don't*

slow us down! Go home if you can't keep up! The sound of their voices
spurred her on to pedal faster, play fiercer, push harder. To get back on
her bike despite the scraped elbows and bloodied knees. Any fuss would
be met with her mom announcing, "You are a Lunt! And Lunts don't
scream!"

Bri's mom, Marianne, was Swedish and had married into a multi-
generational Mormon family. Coming from an immigrant family with
relatives on the other side of the world, Marianne often longed for a big
extended family, filled with lots of cousins, laughter, and love. She was
utterly devoted to her husband and children, always doing her best to
be what their community considered the perfect wife and mother. She
ensured she was at home when the children walked themselves back from
school as even young kids did then; and kept the home spotless and inviting for all who either lived in or visited it.

The big family home was in the middle of rural Utah but was very
much influenced by Swedish architecture. Marianne had had a lot of input
in the design of the house; she wanted to have it focused on the position
of the sun so that every room could be filled with light. She was a firm
believer in the body and mind needing bright sunshine and wide-open
windows, so the house had a huge sunroom with windows that were two
stories high. The family would often gather together in the sunroom during storms to listen to the thunderclap, and watch the desert ignite with
a spectacular show of electricity in the air colliding with the landscape.
The importance of understanding and nurturing the human's place with
the natural world was not lost on Marianne, and she ensured this was an
aspect of life for each of her five children.

Bri's dad, Ray, had built the house with his own hands. He was a
Mormon bishop, and well-respected in their community. In the midst of
all the fearlessness and freedom, the children were also raised to have faith
and to be of service to others; always aware that there were people in the
world who were not as fortunate as them. It wasn't considered appropriate
to dwell on emotions in the Lunt house when there were so many others
in the world who were struggling with far bigger things. The children were
all expected to keep their heads high and their spirits higher, to put their
best foot forward, and to be an example to others.

They were all expected to do their best with the talents that God had given them. Being "average" wasn't good enough in the Lunt family, and the kids were expected to excel at sports or be top of the class. They grew up highly competitive; always keen to outdo each other as well as beat whatever personal best they had set before. The older kids were all very academically-minded, whereas Bri loved playing the piano or doing gymnastics—as a little girl, her greatest joy was feeling strength in her body as she threw herself across the beams and bars, her competitiveness and four older siblings pushing her to do things that other children the same age couldn't do.

Bri knew that she was lucky to be growing up free, healthy, and cared for, with parents who loved her, and siblings who were not only her role models but also her best friends. While talking about feelings may not have been top of the conversation at mealtimes, the whole house was constantly abuzz with the joyful sound of singing, laughter, and plans for the next camping trip or some other adventure they could all have together. It was the American dream.

And then Bri's dad moved out. And everything changed.

Within eighteen months her parents were divorced—all the effort to be the best hadn't been good enough for Bri's dad to stay. Three of Bri's siblings moved out too, including Bri's beloved older sister Trish, who had been like a second mother to her. Little Bri went from being part of what she thought was a happy, busy family of seven, to living with just one brother and a mom who now worked all the hours she could to keep a roof over their heads and food on the table. Instead of her loving mother greeting her when she came home from school, Bri and her brother now let themselves in and waited until mom got home.

Each of the family dealt with the divorce in their own way, which didn't involve talking about it—at least not with each other. In such a small, traditional town with only two other divorces, there were plenty of other people talking about it though, and Bri's school buddies always seemed to be just that little bit too keen to repeat gossip they'd heard from their own parents. Bri dreaded hearing about what may or may not have been the latest development in her family from classmates who seemed to know everything. The taunting continued until Bri learned to respond with some rather colorful language, and the other kids backed off.

The swearing developed into a full-on teenage attitude, along with a love for punk and skateboarding. She was never going to fit in with the popular kids, and she didn't really care. She was happy to be an outsider.

But being an outsider meant a lot of time with her own thoughts. A lot of time wondering if she could have fixed things for her parents. By now her dad had moved twelve hours away, despite her begging him not to go. She'd tried to fix that, panicking at the sight of his car and trailer packed with all his belongings, and throwing herself into his arms crying that she'd do *anything* if only he wouldn't go.

But he *had* gone, and Bri had to adjust to this quieter, less adventurous life in a sadder, emptier home. A home and a life that felt somewhat lacking in love, despite her mother's best efforts.

Both Bri and her brother became rather difficult teenagers, causing all sorts of worry for Marianne who was working so hard. Bri was sneaking out at night, getting into trouble, and dating inappropriate boys who really weren't very nice to her. Whenever she was confronted by her mom, Bri would respond with the attitude, saying that she couldn't *wait* to leave home.

Studying was her way out so Bri pushed herself beyond her limits academically in the same way she could physically. Her determination resulted in her going to college a year earlier than the other kids, where she found the freedom to be herself and the kind of friendship that had been missing in her teenage years.

She also found her future husband.

They were similar in many ways, even down to their names, Brianne and Brian. With Brian Dressen, she found kindness and respect—he wasn't like the toxic boys she had dated when she was younger. Bri didn't just fall in love with Brian, she fell in love with life again, regaining the sense of adventure she'd had growing up—hiking, climbing, and facing seemingly impossible physical challenges in nature together. Mountains formed the stunning backdrop to their courtship—their first date, their first kiss. And with Brian, Bri learned to not just put her happiness in someone else's hands, but quite literally her life too, as they held each others' ropes on their climbing adventures.

With climbing, there was plenty of opportunity to nurture the skills she developed in her childhood cliff-jumping days—where she learned to manage the anxiety she felt before doing something *really* challenging (or really stupid, depending on how you looked at it). She perfected the art of breathing in a way that controlled the rate of her heart, the flow of her blood, and the racing of her mind. This skill would keep Bri pushing through her limits as she entered adulthood—overachieving in most things she'd turn her hand to, including the Bachelor of Science degree in communications that she graduated with at age twenty.

Bri and Brian married within a year of Bri's graduation, and Brian continued studying, eventually building on his chemistry degree to get a PhD, while Bri was the main breadwinner, working as a project manager in the construction industry. They had a decade of adventures before their own kids came along, which Bri raised full-time while Brian started his career with a highly respected but also highly confidential role for the US army.

The Dressens led a more than comfortable life, which enabled them to buy their dream home in a beautiful location, overlooking the mountains upon which their relationship had blossomed. Just like her mother had insisted with her childhood home, Bri insisted that her marital home allow natural light to shine through each and every room, and because of that, there were windows everywhere. It was a stunning house, and Bri was again reminded of the blessings with which life had provided her, trying never to forget her duty to do her best to support others less fortunate.

The Dressens had a son, Cooper, followed by a daughter, Hannah, and Bri devoted her time to being a mother. She became inspired to set up her own preschool, geared toward helping children who struggled to function in a typical educational environment. She painted a bright yellow door at the entrance, and decorated the interior with cheerful colors. The school became a huge hit in the neighborhood, with four classes of eight children each, half of which were taught by Bri herself, and half by another teacher. Teaching in the school felt like the perfect career for Bri—a job that wasn't just about earning an income, but actually provided her with a life that focused on making a real difference to others. Her own kids were thriving, and life was good.

Then Covid happened.

* * *

The Dressens were glued to the news, following all the latest announcements. Bri's sister Trish was an obstetrician-gynecologist in a hospital, so Bri was also getting firsthand reports of what was happening on the frontline, much of which seemed to be about problems with medical supplies. There was a shortage of N95 masks, and Trish was having to make a single mask last a whole day, blood stains and all.

Bri made a post on social media, saying she was looking for people to donate N95 masks. The response took her driving all over the valley, collecting over 1,100 masks in total, and she tried delivering them to the local hospitals. But to Bri's surprise, the hospitals wouldn't allow the donations, saying that they weren't permitted to accept donated masks—only masks that had been supplied through the official channels. *This doesn't make sense. Aren't we in a crisis? Aren't these healthcare workers calling for PPE protection that the hospitals was not providing?*

So Bri got resourceful, and started meeting nurses in hospital parking lots and giving the masks directly to them, then the nurses would sneak the masks into the hospital inventory.

In between sourcing and distributing surgical masks anyway she could, the Dressens did their best to keep the children occupied. They wanted to make lockdown as stress-free as possible for the kids, hoping to teach them both resilience and optimism, community-mindedness, and above all, if the four of them were together, then that was really all that mattered.

They spent hours together as a family, painting animal faces on little rocks, thinking about the people they could cheer up with them. Then they would go around the neighborhood that once was an environment full of kids playing and adults standing outside chatting about the latest gossip. Now it was just a ghost town. Bri and the kids would place the "pet rocks" somewhere on people's front yards, ring doorbells and run away, smiling at the thought of their friends finding these cute little gifts in their yards. Sometimes they would spend evenings attaching glow sticks to their bodies to become life-sized glow-in-the-dark stick figures, then dance in front of people's doorbell cameras. Anything to boost morale in the neighborhood.

Bri was in regular contact with Trish and her kids who lived thirty minutes away—the cousins would make up dance routines to send to each other, or film themselves going to lakes and jumping into freezing cold water, not unlike how their mom and aunt did at Lake Powell all those years ago.

Bri genuinely believed that she was contributing to the solution by supporting the healthcare workers and keeping spirits high, and she encouraged her kids to do the same.

Bri was always wondering . . . what else could she do to help?

* * *

The Dressens were fascinated by the reporting of the vaccine trials. They were both believers in scientific advancements being good and necessary for individual and global health. They and their kids had had all the vaccines that the government recommended.

However, they had decided that the only exception to that would be the HPV vaccine. Many years earlier, Bri's neighbor had shared the first-hand account of the adverse reaction that her nephew had suffered in response to the HPV vaccine. The boy had been left permanently disabled with respiratory paralysis. Out of concern for the many kids in the area, the boy's aunt had urged parents not to give their children that shot. Bri had responded by defending vaccines in general, commenting that not all vaccines were bad, but even so, she had decided that her own kids wouldn't be getting the HPV shot. It wasn't worth the risk.

The great scientific advancement of the Covid vaccine's "Operation Warp Speed" was regularly being reported on in the media, with headlines urging the public to "follow the science." With Brian's scientific background and military career, the Dressens did trust the science, and believed that the world was filled with scientists just like Brian—scientists who would be carefully monitoring the clinical trials, and relaying their analyses to trusted media sources, who were then relaying the information to the public. Like many Americans, the Dressens longed for the day when a Covid vaccine would be made available, and life could return to normal for everyone.

That day seemed to be coming sooner than expected.

By the end of 2019, scientists were reported to have isolated the virus and were making its genetic sequence available to other scientists in

January of the following year. By the summer, pharmaceutical companies throughout the world were reported to be receiving funding from governments globally—as well as foundations and private donors—in order to speed up the development process of a Covid vaccine, with much of the research involving the United States.

US-based Moderna, Pfizer, Novovax, and Johnson & Johnson were all involved in developing products with national research institutes, internationally recognized foundations, or foreign corporations. America as a nation was playing its part in helping to save the world from this crisis, and for some, the sense of national pride started to outweigh the frustration of social restrictions and a general sense of fear.

The Dressens found it both fascinating and exciting to see the announcements about the global developments going on, in particular about how the clinical trials were being managed. As a couple with a strong interest in and respect for both science and medicine—not to mention a beloved family member in the medical field—they were intrigued with how these particular vaccines were being produced so quickly.

In historical vaccine development, the entire process, from pre-clinical trials to manufacturing, would usually take up to ten years. However, because of all the funding and support being generously made available due to the impact of Covid, and the thousands of people willing to participate in the highly publicized clinical trials, it seemed that pharmaceutical companies were going to be able to produce this vaccine within *months* rather than years. In particular, the regulatory process—review, authorization/approval, and manufacturing—would be dramatically reduced for the Covid vaccines. This final process had historically taken five years for other vaccines, but with all the Covid deaths being reported it just didn't seem that the world *had* five years. Reducing the authorization process to less than one year seemed like an incredible accomplishment that just showed what could be achieved if everyone worked together.

The prestigious Oxford University in the UK was working together with the British-Swedish company AstraZeneca, and planning their Covid vaccine clinical trials in the UK, Brazil, South Africa, and the United States . . . in none other than Salt Lake County, just a forty-minute drive from the Dressens.

Bri was thrilled. It was never a matter of "if" she'd get the Covid vaccine; it was always a matter of "when." Of *course* she would do her duty by getting vaccinated against Covid. And now she could not only *possibly* be given a Covid vaccine, but she could play what she saw as a vital role in the actual development of the vaccine. She had two healthcare worker friends who had participated in the Moderna trials, which they'd reported to be a very straightforward process. There seemed to be no reason not to sign up and every reason *to* sign up.

It was just another way to do the right thing.

* * *

The application form was on Facebook; a very simple form, asking for Bri's name, location, profession, and contact details. A representative called to assess whether she'd be a suitable candidate for the trial. They were on the phone for several hours, during which Bri answered questions about her entire medical history—from birth to present day.

Bri had enjoyed a lifetime of good health with minimal issues. She'd been taking a small daily dose of Armour Thyroid to deal with a minor hypothyroid issue that had been identified years earlier, and there had been an episode of West Nile virus—a Dengue Fever-like condition usually spread by a mosquito-borne virus. She'd had infertility treatment for Cooper, and a couple of surgeries. She shared all of these with the clinical trial representative, none of which they considered to be relevant. Bri had no allergies, autoimmune disease, or underlying health conditions that would exclude anyone from the trial. She was physically and mentally active, and fully engaged in life. She was considered an excellent candidate and expected to be called in soon after.

But the trial didn't follow up. Everything went quiet. And Bri assumed that she hadn't been considered a suitable candidate after all.

Then out of the blue, she was invited to the clinical facility where the trial was taking place. Her appointment was November 4th, 2020—the day after the election.

Most of the buildings in Utah were brand-new, but this one stood out . . . and not in a good way. Bri didn't need her background in construction to see that the clinic was in desperate need of modernization. The interior reflected the exterior, with old furniture and lighting that made everything look gray. Funds generated by hosting a clinical trial could be put to good use here.

Bri had expected the facility to be bustling with the hopeful effort of an entire community keen to do their part to end the pandemic, but instead, there was something very sad and empty about the place. The autumn of 2020 was witness to the height of the pandemic panic in the US, with many people too scared and confused to venture out. Maybe the nervous anticipation of the possible outcome of the previous day's political event was making for a more subdued environment than usual—whatever the outcome, it had been a complicated and unsettling few years, and it was difficult for everybody to feel optimistic about the future.

Brianne held on to the optimism that she had tried to instill in her children during the past twelve months, as she sat in the bare waiting room and studied the twenty-six-page consent form that a nurse had handed her upon entrance to the clinic.

The consent form was very thorough.

It began by saying that participating in the trial was entirely voluntary, that it would go on for about two years, and that she could withdraw at any time. She would be expected to have injections, blood tests, and nasal swabs. The safety and efficacy of the product was still unknown, it had not yet been approved, and no other clinical trials for it had been completed.

Two thirds of the participants would receive the vaccine and one third would receive a placebo, which would be just salt and water. The staff were keen to point out that unlike other Covid vaccine studies, their study didn't use another vaccine as a placebo—they used saline so any possible adverse reactions would be very clear. Neither the participants nor the study staff would know who received the vaccine and who received the placebo. It was a "double-blind study"—although, in the case of a medical emergency it would be possible for the study doctor to ascertain what the patient had received.

The purpose of the trial was to see if the vaccine would stop people from becoming sick with Covid—not if it would prevent them from *catching* it, but if it would prevent people from becoming *sick* with it.

The consent form explained that ten percent of the participants would be further studied to see what kind of immune response they developed, and whether they had any side effects such as a fever or swelling at injection site. This ten percent would be required to complete a side effect diary every day for seven consecutive days post-injection, and be periodically interviewed by staff. The other ninety percent would have a telephone call with study staff seven days post-injection to ask about any side effects, and be monitored once a week for the first year to see if they had become sick with Covid.

The consent form acknowledged that there could be as yet unknown health risks to receiving the vaccine, and also that there were no other vaccines or drugs that had been shown to prevent people from becoming sick with Covid.

At that time, 5,000 people had received at least one dose of the AstraZeneca product; thirty people had received a second. Side effects such as a sore arm, chills, headache, and fatigue were acknowledged; along with a decrease in infection-fighting and/or clot-forming blood cells. However, these side effects were mostly referred to as "mild or moderate" with a few considered "severe." Most of these were stated as to have appeared within forty-eight hours, and resolved within seven days.

Serious reactions included allergic reactions, which trial staff were trained to handle. Guillain-Barré syndrome (GBS) was also mentioned as being a rare reaction, and its symptoms were listed, including muscle weakness and tingling of the limbs or upper body; it could lead to paralysis or death. Bri had never even heard of GBS so assumed that it had to be *really* rare.

The consent form also mentioned vaccines that had been trialed against other forms of coronavirus or respiratory illness in the past, and acknowledged that in some cases illness was worse among those who *had* received a vaccine—leading to death in two cases—when compared with those that had not been vaccinated.

The study would conduct pregnancy tests before administering the injection, requested that participants agreed to use reliable contraception during the trial, and stated that if a participant *did* become pregnant, they would be contacted to see how the pregnancy developed. At this stage AstraZeneca did not know how their product might impact an unborn baby.

Participants would be provided with all vaccines, examinations, and medical care completely free of charge, and they would receive payment for visits and phone calls—$125 per completed study visit and $30 per completed phone call.

Anyone who would go on to have a reaction or serious side effects might have their participation withdrawn by the study doctors, even if they still wanted to participate. Anyone who became ill would have medical treatment arranged or referred by the study doctor, and AstraZeneca had an insurance policy that would cover all costs.

If new information relating to the vaccine came to light, then participants would be asked whether they would like to continue, in which case they would be asked to sign a new consent form.

Bri sat in the waiting room reading everything carefully, until the trial investigator invited her into a private room to go through the consent form together, page by page. The investigator had been in the Moderna trial herself, and Bri was very reassured by her attention to detail. Everything seemed straightforward, although Bri did think it was strange that pre-clinical research had shown that monkeys had had a *worse* Covid infection post-vaccine—the investigator explained that they hadn't had a chance to verify that. There just seemed to be a *chance* that the immune system could be weakened.

Despite that, Bri was reassured by the consent form and impressed with how seriously AstraZeneca seemed to be taking any possible side effects or adverse reactions. The investigator explained that the trial had been paused in the UK due to a possible case of MS developing post-vaccine. Bri was then even more impressed with AstraZeneca, thinking they were *really* taking safety very seriously if they were willing to pause their trials for one person to determine if their injury was caused by the vaccine or not. Bri was relieved to be told that investigations had revealed that the participants had had unknown underlying conditions, so the reactions were because of their bodies, not the products. *People's bodies do weird things sometimes.*

Bri felt that if something unexpected happened, or if something went wrong, the trial clinic and AstraZeneca would be there to take care of her. Not that anything was going to happen.

So she signed the form.

A nurse took blood samples and nasal swabs then went off with the form, leaving Bri in the little room on her own as she watched the election numbers come in on her mobile phone. She sat for what felt like a very long time; the Trump and Biden numbers fluctuating back and forth, back and forth. And then the nurse appeared.

"It looks like we're going to give you your shot today!"

Bri followed the nurse down the corridor as she was led into a different room. After all the months of watching the news, distributing the masks, occupying the kids, uplifting the neighborhood, being a good citizen, and trying to maintain an optimistic outlook, she finally had the chance to *really* make a difference. After all the anticipation for the day when the vaccines would restore normality to life again, here she was . . . part of that solution.

But suddenly she wanted to pause *everything* just for a moment. She hadn't expected to be injected that same day. Suddenly, it was all happening so quickly.

Everything in her body was telling her not to do it. Short, gasping little breaths, the pounding in her heart, the tensing of her muscles, and thoughts racing as she faced what she was about to do. She didn't *have* to go ahead with it. She could change her mind. She could turn around and go back at any moment. She could stop this all happening *right now*. But she didn't. She focused on her breathing . . . taking it deeper and deeper . . . she slowed down her heart, cleared the noise in her head, steadied her stance, and calmly found her center before rolling up her sleeve and taking the shot.

The tingling started within the hour.

CHAPTER 2

THE REACTION

The tingling spread from below Bri's right elbow all the way up to her shoulder, and then over to her left arm. By the time she arrived back home, she'd already assumed she'd had the actual vaccine rather than the saline placebo, and she figured that the tingling was a transient side effect that would soon stop.

It didn't.

The tingling continued throughout dinner that evening, after which her eyes started to blur. The family had been sitting down watching the election news when Bri realized that she couldn't see properly. As she tried to refocus her eyes, she found that she could see two televisions, two news anchors, two sets of votes. There was double of everything around her. And her arms were now stinging on top of the tingling.

Suddenly everything sounded like she'd had two cans put over her ears, muffling everything that was going on around her. Hannah was reading aloud as she sat doing her homework on the kitchen counter, but Bri couldn't hear what her daughter was reading. Everything sounded distorted and confusing to Bri's brain. She staggered off to bed, still telling herself that this was all a normal response to the vaccine, and assuming that after a good night's sleep, she'd feel like her normal self again. It was no big deal.

But she couldn't sleep. The muffled sounds and vision developed into severe head pains and ringing in her ears, accompanied by a fever. *It's OK.*

No big deal. This is good—I'm building my immunity. She dozed in and out of consciousness, stumbling back and forth between the bedroom and the bathroom throughout the night. And the tingling still didn't stop. She sandwiched her arms between the mattress and the weight of her body in an attempt to numb her hands so that she wouldn't feel the incessant stinging and tingling anymore. But it didn't help.

By morning the fever had passed, but all the other symptoms had intensified. New symptoms had emerged overnight—she was having trouble controlling her left leg and the rest of her limbs were becoming weaker and weaker. She logged into the app that the trial clinic had downloaded on to her phone during the obligatory fifteen-minute wait post-injection. This was where she'd been instructed to report any side effects. She went through the list and checked all the ones that applied to her like headache and fever, noting that there wasn't anything especially unusual in the list; nothing like what she was experiencing anyway. But there was an "other" option so she checked that, and the app responded with a prompt to contact the trial clinic. Nobody answered Bri's call, so she struggled on with the day, finding noise and light increasingly difficult to tolerate.

Another restless night, and another morning without any improvement to her symptoms, and without a return call from the trial clinic. Bri called them again, while Brian tried to keep the kids quiet and block out all the light in the house—the house they had specifically bought because of all the windows and the bright sunshine. Thick, dark towels were draped on top of the shades—closing the shades alone wasn't enough but the towels at least brought a little relief.

This time the clinic returned her call, brought her in for an assessment and encouraged her to seek out a neurologist to rule out a possible underlying condition for the issue with her leg and the limb weakness that was progressing elsewhere in her body. The Dressens assumed that Bri's reaction was temporary, and in just a few more days, it would all go away and Bri would be back to normal. *It's just my body being one of those weird ones!*

But in the weeks that followed, everything got worse. And the Dressens didn't just seek out a neurologist, but Bri ended up in and out of emergency departments, as one symptom piled on top of another. Nausea, vision disturbances, more tingling, and extreme sensitivity to sound accompanied the limb weaknesses. The pain was no longer restricted to just the nerve sensations in those limbs but now affected other parts of her body and seemed to be exacerbated by any kind of normal, everyday activity. She couldn't even eat. Her bones, her joints, her teeth, her stomach, her legs were painful—*everything* hurt. If she did force herself to swallow any food then she could keep hardly any of it down. She spent her days and nights in and out of the bathroom and lost twenty pounds within weeks. The only thing that brought any relief at all was ginger tea—everything else was rejected by her body in one way or the other.

This was a far cry from the diet that Bri usually enjoyed, which she knew wasn't the healthiest, but it hadn't caused her any problems either. She had always been so fit and active and with such high metabolism, that she'd never even thought about what she ate before. Lots of carbs—sometimes with meat but admittedly very little fruit and vegetables—had given her all the energy that she'd needed to manage the kids and the school, as well as continue the hiking, climbing, and snowboarding that was such a big part of her life.

Now there was no climbing. Now she was struggling to just walk around the house.

The paresthesia had spread to her legs—her entire body was constantly tingling and now her limbs were becoming weaker and weaker. She recalled the GBS symptoms that had been explained to her when she signed the consent form. *It can't be GBS . . . that's so rare! It would be too much of a coincidence. Wouldn't it? Oh my . . . this could be bad. Really bad.*

Bri's legs weren't working properly and felt like noodles as she wobbled her way around. Then she started feeling the *inside* of her body vibrate; like there was a mobile phone buried deep inside her and it kept buzzing and buzzing away. It was non-stop. *Somebody answer it! Pick up! Pick up! Make it stop!*

Her body screamed and screamed at her to get away from everything—away from food, away from sound, away from light. Every part of her body was sensitive to everything. She couldn't even brush her teeth—the pain was too intense and would just make her vomit.

Any glimpse of natural or artificial light felt as harsh as that intense light glaring right into the center of your eye during an eye exam. Apart from when Brian would drive her to the emergency room, Bri spent days lying weakly under the covers in bed with the shades closed, the towels layered on top of the shades, and wearing thick black sunglasses.

Within days, and with neither the clinical trial nor the drug companies being of any help, Brian started looking up scientific papers about vaccines—how they worked, possible serious side effects, what they could do to the human body, and how to treat them. He quickly identified that Bri's reaction could be something to with her immune system, and that it needed calming down, along with her entire external environment. He called in Bri's mom and sister to help out with the kids during the numerous hospital visits, reluctantly telling them that this had happened within an hour of Bri's vaccine but asking them to keep that to themselves.

Despite what was happening to Bri, the Dressens didn't want to be responsible for anybody questioning whether they should take a vaccine or not. They believed that Bri had done the right thing and everything was going to be OK. They knew Bri was having some kind of strange reaction but assumed that it was just her body being weird, or perhaps it was somehow related to the AstraZeneca product specifically. It had to be very, very rare. Nothing to do with vaccines in general. They very much believed in vaccines for the greater good.

Marianne, Trish, and a neighborhood of supportive families were happy to step in, and a schedule was organized so that the kids were occupied and Bri could lie in the dark, trying to keep still and quiet. She couldn't bear anything moving near or touching her body, and the kids were instructed to stay out of the spare room, where Bri was now sleeping. Nonetheless, little Hannah snuck into the bedroom and curled herself around Bri's weary body. *Mommy, I won't speak. I'll be less wiggly. Please let me lie here next to you. Just for a bit. I'll be good. I miss you.*

But Hannah's tiny body was incapable of holding still for too long, and soon there was a little movement of the arms, a wiggle of the legs, then a clicking of the tongue, before a quiet barrage of thoughts flooded from her mouth about all the things mommy had missed that day in Hannah's life. Each tiny word from her angelic little voice felt like knives to Brianne's brain, which responded by urging Bri to do whatever she had to get away from the sound.

Bri had to text Brian to retrieve their daughter; even the sound of her children whispering and the feeling of them snuggling up to her was too much. The household that revolved around the kids was all of a sudden flipped upside down. Everything now revolved around Mom, who used to be the anchor who kept it all together. Bri told herself that this had to be temporary. *This will all go away any time now.*

Stranger and stranger symptoms were taking over—that no doctors seemed able to explain. The tinnitus was like a freight train but was also accompanied by an excruciatingly loud screeching. The nausea was worse than anything she'd ever experienced during either of her pregnancies. She lost control of her bladder and was appalled when she started randomly wetting herself. The opposite was happening with her bowels—her body seemed to have had forgotten how to defecate— she'd sit on the toilet for what felt like hours, concentrating on her pelvic muscles, *willing* them to cooperate so that she could eliminate what little was in her.

Her heart became erratic at random moments throughout the day and night—alternating between slowing down to an alarmingly low rate that almost made her faint, and unexpectedly speeding up in a way that it never had during all her years of athletic endeavors. Her heart had a mind of its own, alerting doctors who could not provide an explanation for why her previously picture perfect low heart rate and low blood pressure were now both constantly high. *What was happening????!!!!!*

Electrical sensations powered through her body, taking the constant pain up to an unbearable level, but without any identifiable trigger or clue as to what would make it wax or wane. Most days, Bri would be jolted awake feeling like she was being electrocuted from head to toe, and would only find a break if she took one of the benzodiazepines, gabapentin, or pregabalins she'd been prescribed. But the medication then made her completely incoherent. It was a tough choice between waking up to getting drugged up, or existing with the relentless shocks. While the shocks were initially terrifying, signaling to her body to do *anything and everything* to make them stop, Bri was reminded of the electric skies in the high desert. She'd watched them with a thrill as a child with her family and avoided them with caution as an adult when out climbing with her husband. Never in her wildest dreams had she imagined that electrical current to course through her veins.

Brian was working from home during Covid, so he was conveniently on call during Bri's confinement. He'd try to get some of his own work done before the kids woke up, then make everybody breakfast—even for Bri, despite knowing that more often than not she couldn't even think about food. Nonetheless, he'd bring up a tray for his wife and sit next to her as she lay there with her sunglasses and headphones on, occasionally wincing in pain or twitching as her nerves ran riot. Sometimes he would need to get her ready for another hospital appointment he'd arranged, but usually the couple just sat in silence, as tears rolled down Bri's face, soaking the blanket Brian had gently tucked around her.

Even the lightest of touches was overwhelming for Bri, but it felt *good* to have Brian so close, even if it was for only a few minutes. He had always brought her so much comfort during their years together. But the feelings of comfort that Brian's presence brought quickly became mixed up with unsettling ones. His hands were reassuringly familiar, but also felt startlingly new—his face, his voice, his smell. Even the swish of his pants as he tentatively walked from the bedroom door to the bed was alarming to Bri's ears. It was like she had to relearn everything sensory about the man she had been married to for over a decade, and relearn everything sensory about the rest of her life too: how her legs felt when they moved, what food felt like to chew, the feel of fabric as her pajamas brushed against her body. Relearning the tiniest details about life and how her body processed the elements of it.

It was a task so daunting that all Bri could do was return to what had given her the strength to jump off those cliffs with her brothers and Trish, and just *breathe*.

"Am I going to get better?"
"Yes."
"How do you know?"
"I just know. There is *no way* a vaccine could do this."
"You promise?"
"I promise."

Bri asked the same question every morning. And every morning, Brian gave her the same answer. He was researching every spare moment he could, trying to find scientific papers about the long-term adverse reactions to vaccines, and was struggling to find anything. Surely this would eventually go away. Bri *would* get better.

In the meantime, the laundry started piling up. The dishes sat unwashed. The snow settled on the driveway until it slowly melted on its own. And the kids started having their own meltdowns. The household was falling apart.

* * *

"It looks like a bit of a disaster in here."

Trish was not inaccurate in her observation and took it upon herself to hire some help for her little sister.

Bri hadn't told anyone outside of immediate family that her sudden illness had started within minutes of the AstraZeneca vaccine. It felt very wrong to be broadcasting this when so many people seemed to be struggling with Covid-related illness, not to mention the fact that her physician sister was witnessing firsthand the challenges that the Covid situation was responsible for. Trish had been like another mother to Bri, and Bri valued her opinion and respected her not only as an individual but also for her dedication to her profession. Trish herself was deeply distressed at seeing her little sister so badly impacted, let alone by a pharmaceutical product and one that she wholeheartedly believed in at that. Bri's adverse reaction to the vaccine was a topic that the two sisters would have to navigate very carefully as the months went on. But Trish rallied around, despite whatever conflicting feelings she may have had.

Others rallied around too. The Dressens had deep roots within their community, and neighbors scheduled a rotation of home-cooked meals when they heard of Bri's mysterious illness. The preschool parents had jumped in to teach the classes in the early days when everyone—including Bri—thought she'd be back in action within a week or two. As the weeks went on and it became apparent that Bri's illness was not letting up and the doctors didn't know what to do, a substitute teacher was hired, creating a sense of relief for Bri but also adding to the fear she felt about the future. *Is this my life now?*

Just weeks earlier she had been hanging out with eight tiny preschoolers, listening to them talk about dandelions, and crayfish, and why purple is the very best color ever! She had loved hearing them scream out of pure joy at something new they had seen or heard in the little preschool classroom with the yellow door that Bri herself had painted. Now she was confined to the upstairs of her home where she couldn't hear a thing.

Friends wanted to visit. When the Dressens had turned down invitations from friends to join in the usual holiday traditions that Christmas, Brian had called a few of them individually. He'd sensitively explained that Bri had had an adverse reaction after participating in a clinical trial; they'd been kind in their responses and keen to visit, assuming it would be like visiting a friend recuperating from surgery. They were completely unprepared for what they found. Who had ever heard of an adverse reaction to a vaccine?

One friend, Jenny, drove two and a half hours to see Bri, only to audibly gasp when she did. The sight of her formerly striking friend shuffling into the living room in her pajamas, wearing sunglasses and headphones, was enough of a shock, but she couldn't contain herself when she hugged Bri, who was now mostly skin and bone, quivering and very frail. Bri frantically told Jenny that if she could just find a way to sleep and eat then she could survive, and Jenny burst into tears. She cried for the entire journey back home.

Other friends visited, and because Bri couldn't cope with them talking to or holding her, they sat in the darkness in a corner of her room so she at least knew she wasn't alone in her suffering. They didn't push her to talk about what had happened, and what was still happening, because they could see that Bri was overwhelmed with pain as she raised her hand to indicate that they should leave. Everything was too much.

A cousin shaved Bri's legs as she lay there—a simple yet complicated act that showed such kindness as she so very, very gently drew the razor down Bri's sensitive legs. Such a moment of intimacy and vulnerability that the old Bri would never have allowed herself—having always been the helper of others, she was now experiencing what it felt like to be on the other side of such humanity.

When the cousin left, she'd placed a little letter next to the bed, for when Bri wanted something to read. Reading was such a struggle with the ongoing blurry double vision and difficulties processing information, so the letter sat there for days until Trish visited and offered to read it for Bri. She opened the envelope and started to read out loud, *"Dear Bri . . . I can't imagine how hard these last few months have been. I have felt helpless in this situation and often want to reach out and help but it feels like there isn't anything I can do to take this away. But I want you to know there is ONE person who can, and that is Jesus Christ . . ."*

Trish's initially confident voice started trailing off as she continued to read. By the end of the second paragraph, she stopped reading out loud altogether and read the rest of it to herself.

Bri and Brian, while both raised Mormon, had begun to doubt their religion during the early years of their marriage. Like many young people raised in any devout environment, the Dressens had gone through a period of uncertainty about their church—a church that they had both dutifully followed until they felt they had reason to do otherwise. And after years of research and reflection, they decided together to move more and more into the background of their local church. They still lived by the same values with which they had been raised, were raising their children in the same way, and respected the community within which they lived and to which they contributed, but they no longer considered themselves Mormons.

The letter, while genuinely written with love and kindness, suggested both directly and indirectly, that Bri's sudden illness was because of her lack of activity within Mormonism. Trish finished reading the letter silently to herself, folded the pages, put them back in the envelope and, in her usual decisive way, announced that now was *not* the time for Bri to be reading this. Bri was just too fragile, and Trish wanted Bri's promise that she wouldn't be picking up that letter anytime soon. Bri always followed her sister's orders, even as an adult, and she assured Trish that she wouldn't touch the letter.

Bri often found herself reassuring her visitors—their shock at Bri's transformation was visible, and she didn't want everybody worrying about her. So she'd somehow find the ability to speak and tell them that it was all going to be OK even though she really wasn't sure if she believed that herself anymore. No doctor had been able to tell her that, after all.

She sat listening to her well-meaning friends' stories about what was going on in their lives as they attempted to take her mind off whatever was happening to her body. But not only was it an unbearable effort to *physically* listen and process what they were saying, it was emotionally devastating. Bri couldn't bear hearing about all the fun things they were doing with their other friends, their partners, their kids, *their* lives. She was happy for them but simultaneously reminded of a life going on out there that she was no longer part of.

Bri's mom visited nearly every day. She had had a stroke many years before but was mobile, physically capable, and just as determined as she'd ever been to provide for her daughter. Bri hadn't given her much opportunity to do so over the years. Their relationship had floundered after Bri's teens when she had been so "difficult." It was heartbreaking for Marianne to see her previously feisty daughter mostly bedridden, struggling to walk, and so very sad at being unable to share in her own children's lives.

Marianne stepped in with the kids—making their lunches, collecting them from school, dropping them off at their friends' birthday parties, and trying as best she could to maintain some sense of normality for them. But the children felt the absence of their mother in Nana's sandwiches; Nana always forgot the special peanut butter.

And Marianne stepped in with Bri too, tentatively at first, sitting at the end of the bed. Not too close. Not too loud. She would squeeze Bri's feet one at a time, as if she was walking—left, right, left, right—and there was something about the squeezing that made Bri feel like a real person again. The pain would ease just enough to make Bri feel safe, and that was when the tears would flow.

As Bri cried, having her feet gently rubbed by her mother, their relationship found a peace that had eluded them for many years.

Marianne softly told stories of her own childhood; stories that Bri had never heard before. Stories of the grandmother that Bri had stayed with who, widowed young, had made her way from Sweden to America in the hope of a better life for her two young daughters. Stories of the beauty of Sweden through a child's eyes—a child who delighted in the joys that the seasons would bring. Long, cold, dark winters spent huddled up indoors yet always tempered with the anticipation of spring being just around the corner. The first daffodils of the year would signify that the dark days were over, and life was about to begin again.

Sitting in the dark, listening to the sound of her mother's voice through her earplugs, and feeling the touch of her mother's hands, Bri felt just a little bit of hope wash over her. Marianne took Bri out of her own head, even for just a short time, and her body and her spirit found just a little respite. *Maybe it would all be OK, after all?*

Bri decided to focus on learning to walk properly again. But first she had to do something about the "leg noodles." Her body had deteriorated to the point where, aside from the hospital trips, she was now bedbound, and her formerly toned and muscular legs rapidly lost their shape and strength. Being a longtime active athlete, Bri knew a thing or two about recovering from sports injuries, and she understood that not using a limb after injury was one of the worst things someone could do. She had been aware of this as soon as she had become bedbound; and during those months tried to force herself to do any exercises that a doctor suggested. If she could just *move* her legs then perhaps all the signals inside her body would start to make sense, and walking would become second nature again. If that didn't work, then she would just work harder and harder until it did; harder than she ever had before.

She had shuffled down the hallway to the spare room where the spin bike sat neglected then leaned on the bike, exhausted from the effort it had taken just to make it down the hallway. Focusing on her breathing, she rested her butt on the seat, and lifted one leg over the bar, holding herself upright as she leaned on the handlebars. More breathing. She placed one foot on one pedal, then the other foot on the other pedal. And willed her legs to just *push*. Just the way she'd had to concentrate on telling her body how to go to the toilet, she was now concentrating on telling her body how to ride a bike again. Her upper legs moved—slowly and painfully— but they moved. Her feet fell off the pedals, but she would catch herself, recalibrate, and try again. But the harder she tried, the more her body felt like it was falling apart. Any simple movement caused the vibrations within her body to go into overdrive. It was too much. Tears of frustration poured down her face as she remembered how she had spent hours on that bike without even thinking twice, pushing her body beyond its limits, just like she'd always done. Now she couldn't even push the pedals.

All she could do was breathe.

So that's what she did. Focused on breathing. All day. Every day. Shut the world out with sunglasses and earplugs. Tried not to think about what life was like before. Tried not to think about everything her body and mind used to be capable of. All she needed to do was *breathe*. One breath in. One breath out. One breath at a time.

* * *

February 2021—four months after the vaccine—was when the earplugs came out. And Bri learned to walk steadily again rather than shuffle. She had always been able to stand, but her little shuffle had triggered such disturbing sensations throughout her entire body that it had become too disheartening to continue trying. Instead, she found that she was able to *very gently* nudge her feet out . . . not too far and not too quickly . . . just a very gentle flick of the toes back and forth. So she carried on doing that while lying in bed. Nudging her feet backward and forward, backward and forward. Feeling the muscles in her calves and thighs cooperate with this delicate movement encouraged her to keep going. The muscles were there. Her body could do this. She just had to just keep going.

After a month of gently moving her feet, Bri was outside—not quite walking but definitely doing a little more than shuffling, as she moved forward on the driveway together with her mother. Marianne's stroke had left her just as unsteady, and the two women wobbled along together, arm in arm, giggling at the thought of what a sight they must look. Bri had struggled to find any joy in life, but in that moment, she found herself able to laugh again.

If I can breathe, I can make it.
If I can sleep, I can make it.
If I can eat, I can make it.
If I can walk, I can make it.
If I can find joy again, I can make it.

Joy was understandably lacking in Bri's life. Her condition had removed her entirely from all the things that brought her joy—her family, her friends, her work, and her adventures. Whenever she'd been stressed or upset in the past, she'd always had a number of activities in which she could immerse herself—they were now all off-limits. She had been put on antidepressants, and almost immediately had a reaction to those, feeling her entire body burn up from the inside. Any step forward seemed to be accompanied by a big step back.

With every tiny improvement in Bri's symptoms, came an increased general frustration at her new life as she realized just how limited that new life was. As the weeks and months went by, the likelihood of Bri ever returning to work began to wane.

"How do you know I'm going to get better?"
"I just know. I'm going to fix this. I'm going to sort it out."
"Do you promise?"
"I promise."

Brian's determination to figure out what was happening to his wife had led him to be in regular contact with numerous doctors, researchers, and government bodies. Despite all the reassurances in the trial consent form, he had realized that he was going to have to figure out what to do on his own.

The experts had no idea.

CHAPTER 3

THE EXPERTS

After Bri's second call to the trial clinic about her strange symptoms, they had asked her to come in for an evaluation, telling her that what she was experiencing, was "not normal."

The Dressens drove to the test clinic forty minutes away, with Bri juggling a mixture of emotions along with her distressing symptoms. *What does "not normal" mean?* The phrase kept repeating itself in her head. She told herself that it was just her body being weird because *surely* vaccines couldn't do something like this. And even if it was somehow related to the vaccine, then she was doing her part by reporting it so quickly, by going in for an evaluation as soon as the clinic asked her to, and by supporting the experts as they conducted a full investigation. She was fulfilling her obligations as a trial participant so that AstraZeneca could respond appropriately as they continued to develop their product. The consent form had made it very clear that AstraZeneca valued the trial participants in their promise to provide all the medical care necessary if it was required. Bri reassured herself that everything was going to be OK. This was all just part of the process.

The principal investigator of the trial conducted the evaluation herself, along with a nurse, checking Bri's blood pressure, heart rate, and basic neurological functioning. Bri had met the investigator before—she was the member of the trial team who had gone through the consent form's hard copy page by page in great detail two days earlier and been very reassuring. Bri remembered that the investigator had mentioned that she had participated in a trial herself—the Moderna one. Bri felt safe in her hands.

The investigator told Bri that she might have multiple sclerosis and to go to a neurologist to get tested. MS was one of the conditions that had been identified in someone in the UK trials, and their post-injection symptoms had been one of the reasons why the trial had been put on hold. *Just someone's body doing something weird. Nothing to do with the vaccine.*

Maybe her own body was doing something weird. The investigator said she would contact AstraZeneca and get their input. If it turned out *not* to be MS then they'd have to put the trial on hold.

Multiple sclerosis would explain a lot of Bri's symptoms—especially the tingling and vision problems—and the Dressens were keen to get Bri in front of a specialist as soon as possible, but the trial clinic hadn't introduced or recommended anyone to them. They were at a loss about where to go, but wanted to be as cooperative as possible, feeling responsibility toward the trial and knowing how important it was to get Bri's diagnosis to them. The clinic staff said that Bri had to be really sure it was just her body and not the vaccine because there would be ramifications if there was *any* possibility that the vaccine may be responsible for Bri's symptoms. It seemed that the future of the trial and possibly even the AstraZeneca vaccine as a whole might be impacted by Bri's reaction—it felt like a huge burden, but the Dressens stepped up. Brian set to work calling all the neurologists in the area.

The local neurologists were fully booked for months but they all said the same thing; that Bri really ought to go to an emergency department, especially as the tingling was spreading throughout her body, and the other symptoms Brian described were getting worse.

So Bri had checked in to ER, two days after the vaccine. This was to be the first of *many* hospital visits.

Brian hadn't been allowed in the hospital and Bri was sent to the Covid section of the ER unit, despite her assurances to the staff that her symptoms weren't because of Covid. The room was packed full of people with what looked like Covid symptoms, some of which weren't unlike Bri's own symptoms, and she wondered if any other people sitting there had also participated in the trial. The effects of the vaccine and virus looked so very similar.

But at that time, everything was about Covid.

Bri kept correcting staff who assumed she had Covid, informing them that she'd been part of a clinical trial for the vaccines, but it didn't seem to register with anyone—people seemed to hear the word "Covid" and stop listening after that. Nobody had heard of anyone being in a clinical trial, and it wasn't until they'd confirmed Bri's Covid test as negative that the obviously busy staff started paying attention. The whole department was just too overwhelmed. And besides, the hospital staff assumed that a trial participant would be dealt with by the trial medical team. Or the drug company itself.

Bri explained that she had to get conditions ruled out so the trial could decide what to do next. Everybody seemed to understand the importance of an accurate diagnosis, and Bri was put in a private room while one doctor after another came in to talk about her symptoms. But then she noticed that there was a lock on the door which the medical team used every time they came in and out of the room. *Do they think I'm crazy? Are they going to tell me this is all in my head?*

The hospital was bright and the overhead lights were painful to her eyes, so she switched them off and lay in darkness trying to calm her mind, telling herself that this was all going to be worth it. *This is what "doing your part" means.* And just when she'd calm her thoughts, another doctor would walk in, turning the lights on, and the excruciating eye pain would start again.

Being taken down to the MRI machine in a wheelchair was a non-stop barrage of visual and aural stimulation. Doors opening and closing, people talking, machines bleeping, simply passing by pictures and posters hung in the hallways—all the normal sights and sounds of a hospital—merged into an onslaught of sensory information that Bri's brain couldn't cope with. The MRI was actually a relief despite the sound—at least her body wasn't moving, and she could wear an eye mask inside it.

ER didn't find anything significant, and Bri was diagnosed with a "vaccine reaction" and prescribed a small run of steroids to calm down the immune response and inflammation. She'd been in the hospital for the entire day, and it was pitch-black when she was finally discharged. As she slowly made her way out into the fresh air, she was soothed by the relief that the darkness brought to her sore eyes.

The ER doctors had been kind and sympathetic but admitted that there was nothing else that they could do, and told Bri to report back to the trial clinic so that the drug company could get involved. Bri did report back and assumed that the official recognition of the vaccine's role in her symptoms would be invaluable to AstraZeneca. The clinic seemed to appreciate the feedback and urged Bri to continue trying to find out what was going on with her body. They told her not to worry about the cost, which would be reimbursed later. They would feed everything back to AstraZeneca, who would be in touch soon.

Bri waited for AstraZeneca's call, sure that the pharmaceutical company would be keen to find out all the details, and the Dressens kept Bri's ordeal to themselves. It didn't occur to them to speak to anyone outside of the trial team or to medical professionals, and they continued to go through the proper channels as they persevered with their quest for answers. It was important that they got as much information as possible not just for Bri's health but to support the trial. They kept quiet because they didn't want to create unnecessary worry—after all, this was exactly how vaccines were made safe.

Brian kept pushing for the MS testing, knowing how relevant that was to the trial, and finally managed to get Bri a series of tests and consultations, which eventually ruled out multiple sclerosis, as well as the other neurological conditions that he'd found mentioned in scientific papers as occurring post-vaccination: transverse myelitis (TM), Guillain-Barré syndrome (GBS), and acute disseminated encephalomyelitis (ADEM). It was also confirmed that Bri hadn't had a stroke, and the West Nile virus that she had had many years before hadn't been reactivated.

GBS had been mentioned in the consent form as a very rare but potentially paralyzing illness causing weakness and tingling, and possibly death. TM and ADEM were conditions relating to inflammation of the spinal cord or central nervous system, sometimes leading to permanent damage or death. The Dressens were relieved that Bri didn't have any of these conditions, but apprehensive for what this would mean for the trial, which would surely be put on hold now.

But the trial wasn't put on hold.

Bri's symptoms were getting increasingly worse to the point where the trial became less of a priority for the Dressens. Steroids hadn't helped at all. Life became a constant round of trips to one hospital or another punctuated by visits to ER—at one point stopping off at ER *on the way* to a hospital appointment.

Each hospital visit followed the same routine—it was assumed that Bri had a Covid infection as opposed to a Covid vaccine so Brian was not allowed in, and Bri was put in the Covid unit until multiple Covid tests would come back negative, at which point she would be moved to a non-Covid part of the hospital, where Brian was then allowed to join her. The medical staff would express interest in the fact that she'd had a vaccine as part of a clinical trial and want to investigate a little, but never seemed to find anything of significance or that they were able to assist with. So the inevitable, "What's AstraZeneca saying? What's the trial company saying?" questions would be asked, to which the Dressens didn't really have a response. The doctors were frustrated because—after ruling out anything else—they concurred that the vaccine was the most likely cause of Bri's debilitating symptoms, but they had no idea how to treat her. Even a neurologist who told her that he'd seen something similar in the past after other vaccines had no idea what to do about it. The medical profession needed guidance from somewhere—these vaccines were about to be rolled out to the public.

During her multiple trips to hospitals, Bri became acutely aware of just how much people were depending on these vaccines. The Covid wings were terrifying places; one had staff walking around in sealed protective gear that made them look like astronauts, and patients were all housed in domes outside of the hospital itself complete with their own air filtration system. Bri was sent from one extreme to the other when they moved her to a huge room with rows and rows of beds, all set up like an old hospital from the 1900s. She lay on a bed next to someone who was coughing up blood into a cup that was already almost full.

The entire hospital experience confirmed everything that Bri had seen on the news—people were really, really sick, and trauma level one emergency rooms were at breaking point. She was very aware that her presence in the hospitals was putting additional pressure on the health system, which brought about so many conflicting feelings; she'd taken this vaccine so as not to be a burden, however here she was being *very much* a burden *because* she took a vaccine. A pharmaceutical product, designed to prevent sickness, had resulted in her becoming very, very sick. It was a confusing thought, compounded by the realization that doctors had absolutely *no idea* what to do if someone had this kind of reaction to a vaccine . . . not just the Covid vaccine but *any* vaccine, so it seemed. There was nothing for them to go on, whereas billions of vaccines had been administered in the past—surely there had been reactions like this before? What happened to those people?

In the absence of any pathways or protocols that the doctors could follow, they tried treating Bri symptomatically. She was put on gabapentin for the nerve pain, which only made her feel worse with side effects that especially affected her cognitive functions. On another visit, she was given a high dose of anti-dizziness medication, which stopped the nausea but turned her into a zombie—the doctor had to shake her awake to get any sense out of her, but the nausea had gone, so the doctor had done their job. Bri was pronounced well again and sent home, barely able to keep her eyes open from the medication.

At home, her heart rate shot through the roof, her feet went numb, and then she lost all feeling in her legs. Brian made calls to the clinic, and to the ER that they just visited, but nobody wanted to know. He gave up and lay next to Bri for the entire night, with his hand gently resting on her chest, reassuring her that he wasn't going anywhere, that she was going to be OK, and that he would get to the bottom of it. But the reality was, Brian knew that his wife's heart wasn't beating in a way that was anything close to normal, and he couldn't bring himself to take his hand away. He was afraid her heart would stop beating altogether.

The next day, he called the neurologists who had ruled out MS, only to be told that they had no idea what to suggest other than emergency, so it was back to ER again where all the tests indicated severe problems with Bri's heart, blood pressure, and neurological functioning. *What's causing this now? Is this the vaccine?* The doctors couldn't identify what exactly was going on or why, so told Bri she could go home. Brian was incredulous,

"She can't move her *legs* properly and you're going to send her home?!" They reluctantly admitted Bri but made it clear with lots of eye-rolling and sideways glances that they thought this was a psychiatric issue. None of the known tests were coming up with anything after all, so the only other explanation had to be her mental health.

Bri's mental health understandably started to be affected by her physical health and the complete disruption to every single element of her life, so she didn't mind when she was referred to a mental health specialist—maybe they'd be able to offer some kind of treatment. The psychiatrist acknowledged that she was justifiably upset because her body wasn't working properly, but the reason her body didn't work was *not* psychiatric. He didn't know what to do but mental health treatment was not the answer.

Nonetheless, Bri was put on one form of antidepressant or benzodiazepine or another on top of all the other medications she was given; all of which only seemed to make her feel worse both physically and mentally. It was as if her entire body had become hypersensitive to any kind of pharmaceutical intervention.

Bri had no idea how to manage all of these new medications, none of which seemed to get to the root of the problem, and nobody offered her any guidance—she had no psychiatric history and had never needed any of these kinds of drugs before. She'd gone from being healthy and thriving to taking fourteen different medications a day.

When Bri visited her primary care physician for the first time, the doctor had no idea what to do either. She had been Bri's doctor for five or six years, although only saw her when she went for her annual checkup, at which Bri would always receive a clean bill of health. Bri got on well with her; the two women were both avid skiers and would often talk about the best deals to be found when planning their ski trips. Like the hospital doctors, Bri's primary care physician had no idea what to do either; she wanted to help but didn't know how, so just asked the same question that everyone else did, "Is the drug company doing anything?" And urged Bri to get more forceful with AstraZeneca to provide support.

Bri was exhausted. She had nothing in her to be forceful with anyone. The boundless energy she'd always had at her disposal was nowhere to be found. Dealing with the entire health system was a medical merry-go-round. Nothing actually seemed to have been put in place to deal with anyone who had an unexpected reaction to the vaccine. Bri assumed she was very, very rare—one in a million maybe—but still, wasn't that part of the purpose of a trial? To find that one in a million, to understand why, and to then treat? Especially for something that would be administered *to* millions?

Exactly two weeks after the vaccine, the trial clinic asked Bri to come in for a second evaluation. The Dressens drove to back to the clinic, Bri in so much pain but desperately holding on to the hope that at last, something was happening; *somebody* was going to do something.

As soon as Bri entered the trial clinic, she was directed to the Covid wing rather than the clinical trial area in which she'd had the first assessment. She was going to be seen by a different team. This team were all wearing the "space suit" protective gear that she'd now become familiar with, and were far less friendly than the actual trial team had been. It was quite a different experience than her first two visits had been—no warm welcome or reassurance this time. *Maybe they're just afraid of Covid. They must have such a difficult job.*

An administrative staff member told her that nobody could see her until she'd signed a new consent form. Bri asked what was different about the form but couldn't get anyone to explain it to her; they were too busy and just kept telling her that they couldn't help her in any way until she signed it. When she signed the first consent form, somebody had patiently gone through every single page of it; this time was completely different.

"Please, please help me."

"We can't do a single thing to help you until you sign this."

A laptop was placed in front of her and the employee left the room, saying that she would return after the form was signed. Bri couldn't even see properly to be able to read it. She still had trouble with her vision. Even if she *could* see properly, the tears pouring from her eyes would have made it impossible. She had no idea what the long and cumbersome amended consent form said but felt she had no choice but to sign it. Otherwise, they wouldn't help her, and she needed help. She signed it. The clinic staff entered the room, performed two nasal swabs and a blood test, then told Bri she could go. Nobody asked her anything about all the symptoms she had been dealing with since one hour after getting the vaccine. The evaluation was only to determine whether Bri had Covid or not.

She cried all the way home.

The Dressens were utterly confused and couldn't understand why the clinic wasn't interested in doing anything more than several Covid tests, which unsurprisingly came back negative . . . *again*. Had the clinic just wanted her in there to sign the form? Where was the Oxford team the clinic kept saying were going to be in touch? The "excellent team of neurologists" that AstraZeneca apparently had? Where were all these enthusiastic scientists and medical professionals interested in trying to get to the bottom of what was going on? She couldn't understand why the test clinic kept promising that AstraZeneca would be dealing with it "any day now." But nothing. *This is their product . . . why don't they WANT to talk to us about this? What do we need to do to get help here?*

Brian started looking at other options, and reached out to the CDC. No response. He filed a VAERS report, as suggested by both the CDC and FDA websites, and didn't hear anything back after that either. *Was literally nobody interested in this?*

Within a few weeks of vaccination, Bri's health deteriorated to the point where she wasn't even thinking about AstraZeneca anymore. Brian took over all the communication with the trial team, keeping them posted on every hospital visit. The trial clinic staff continued to be polite and encouraging, telling him to go ahead and schedule more appointments and get more tests; and not to worry about the cost, which would be reimbursed as per the terms of trial participation. So Brian kept sending over expenses and receipts, and the trial clinic assured Brian that AstraZeneca would be in touch soon.

Brian looked into contacting AstraZeneca directly but struggled to find a phone number or an email address that wasn't just a general customer service contact. He couldn't find any guidance on their website regarding what to do or whom to contact in the event of an adverse reaction to one of their products. It was a huge multinational corporation with many locations, and even contacting the US branch seemed impossible. Bri had been given the contact information for the trial clinic, and always told to go through them rather than contact AstraZeneca directly.

It was like the trial clinic provided a protective buffer that made it very difficult to communicate with anyone who might actually be able to do something. And there was no actual proof that the trial clinic was passing on everything that Brian was reporting to them. It was very frustrating and confusing.

By this time, if Bri did speak to the trial clinic herself, it was only to beg them for help, with tears pouring down her face. She just needed somebody to help her and surely it would all go away? Whatever this was, it would stop, right? Brian agreed with her, still assuring her that vaccines couldn't possibly do this. There must be something else underneath it all. If only they could get someone to look into it. So they held on, thinking that even if they couldn't get anyone to help, at some point it would just resolve by itself, wouldn't it?

In between dealing with the children, the house, and his own job, Brian was still reading all the scientific papers and articles, trying to get as much information as possible about what might be happening to his wife. His doctorate in chemistry and his scientific background meant that he could quickly and easily understand the published papers, and within days of Bri's reaction he had come across a treatment called IVIG—Intravenous Immunoglobulin.

IVIG is a blood-derived treatment that is slowly injected into the body of people whose immune systems are struggling to function. Brian noted that it was used both nationally and internationally in the treatment of some of the conditions that were, at that point, possible explanations for Bri's symptoms. He found a neurologist in Germany, Prof. Dr. Harald Prüß, who agreed to do a blood test, so set about arranging the logistics of getting a centrifuged vial of Bri's blood to Europe. The results confirmed that Bri had anti-neuronal antibodies, indicating that her immune system was attacking her nerves and affecting the way that her nervous system was functioning. He recommended a drug usually used in cancer treatment called rituximab, a kind of blood "washing" treatment called plasmapheresis, or IVIG. Brian printed out copies of the research papers as well as Dr. Prüß's recommendations to show the doctors and other medical experts that were baffled by Bri's condition. He begged them to try Bri on IVIG. They ignored him.

Brian kept going, sending email after email to immune system researchers or government bodies about Bri's symptoms, mentioning that the symptoms had started very soon after the vaccine, and asking for help. He assumed that, given the importance of the Covid vaccine, there would be plenty of interest in what had happened to a participant of an ongoing clinical trial. But the few responses he got back all said the same thing; ask the drug company to deal with it.

But still, AstraZeneca weren't stepping in. The only indication that AstraZeneca were even aware of what was happening to Bri was when they "unblinded" her at the request of an ER doctor who wanted confirmation as to whether Bri had had the saline placebo or the vaccine. Bri was pretty sure she hadn't had the placebo, and AstraZeneca confirmed that fact. They also said that Bri would not be given a second dose even if she'd wanted it, but they still wanted her data. They said nothing about how she should deal with the ongoing deterioration in health.

Brian came across papers about something called a "spike protein"—a protein that was on the surface of all coronaviruses, not just Covid. It was highly toxic and considered to be the reason why Covid itself was making some people so ill. This spike protein seemed to have played a significant role in the development of the Covid vaccines, which Brian then learned worked in quite a different way to how other vaccines worked. Whereas other vaccines work because the body develops antibodies to what is injected, the AstraZeneca vaccine injected "instructions" for the body to make this spike protein, and then the body developed antibodies to the spike protein, not to the virus itself. And it wasn't just the AstraZeneca vaccine that was using this method; other Covid vaccine manufacturers were using it too. It was a completely different kind of vaccine with a new way for the body to develop immunity. It had never been used before, but it had been in development for decades. If there was a problem in teaching the human body to create such a toxic substance as the spike protein, then there would be plenty of research on it. *Was it possible that the vaccines could do the same as the virus?*

Brian continued contacting specialists in neurological conditions, in the back of his mind wondering not just about the AstraZeneca product, but about all the other vaccines in development using the same technology.

He had gotten used to receiving the usual response—"What's the drug company doing?"—if he got any response at all, and didn't expect to hear anything from the National Institutes of Health (NIH), so was surprised when Dr. Avindra Nath replied straight away. Dr. Nath was a specialist in neuroimmunology—how the nervous system and the immune system interact—and the director of the National Institute of Neurological Disorders and Stroke at the NIH. He was a researcher at the National Institute of Allergy and Infectious Diseases and worked directly under Dr. Anthony Fauci, who advised President Donald Trump on policies during the pandemic, and was about to be appointed chief medical advisor to President Joe Biden. Dr. Nath was highly respected in his field, and Brian was encouraged by the fact that he seemed keen to speak to the Dressens, scheduling an online appointment within days.

Bri sat curled up in a ball on the couch under a blanket, with her head on Brian's shoulder, as Brian did most of the talking in front of their laptop. Dr. Nath and his colleague Dr. Farinaz Safavi were intrigued by everything that the Dressens had been through—the trial procedures, the clinic's response, AstraZeneca's lack of contact, and the doctors' treatment of Bri during the many hospital visits. Bri's medical records made it clear that there was no doubt that the vaccine had caused her ill-health: "vaccine reaction," "vaccine side effect," "immunization reaction," "likely side effect due to an increased immune response to the vaccine," "post-vaccine polyneuritis," and "Covid vaccine injury." The confirmation was scattered throughout her notes.

Dr. Nath didn't question the cause of Bri's illness and was full of compassion for the couple. He even seemed sad when Bri said that she was one of the first people to be injected after AstraZeneca was allowed to resume their clinical trials in the US. Dr. Nath explained that he and Dr. Fauci had debated whether to allow the trials to continue or not after multiple sclerosis and transverse myelitis were diagnosed in a couple of the participants in the UK trial, "We were wondering if the problems in the UK would emerge here."

Dr. Nath gave the impression that looking into adverse reactions wasn't really what the NIH did, but he was keen to help nonetheless, offering to speak to neurologists, provide referrals, and even consider starting a study, with Bri as the first patient enrolled. The Dressens were so appreciative of Dr. Nath's understanding and willing to go beyond the NIH remit in order to help Bri. *Finally! Somebody is on our side!*

But Bri's relief was short-lived. She was just too exhausted. It had been almost two months of fighting to be heard, and her health was still deteriorating. Amid much fanfare, Pfizer's Covid vaccine had just been authorized in the United States, and Bri had become tormented with thoughts of regret. *What if I'd just waited a few weeks? What if I'd gotten another vaccine? Maybe I wouldn't be dying.*

The trial clinic stopped asking Bri for her data, stopped encouraging her to go for more tests, and stopped assuring her that AstraZeneca would be in touch soon. *They stopped acting like they cared.* And on the other side of the Atlantic Ocean, on December 30th, the MHRA (the UK's equivalent to the FDA) authorized AstraZeneca's Covid vaccine for use throughout the UK.

While AstraZeneca was celebrating, Bri was suicidal.

CHAPTER 4

THE DARKNESS

Bri's head and eye pains may have been alleviated by blocking out the light, but it wasn't long before the darkness wasn't just surrounding her physical being, it started to seep into her soul too. Maybe her mom had been right all those times when she'd insisted on throwing open the windows and shutters in the family home, declaring, "let the light into your heart to flush out the darkness!"

There was no light in the house, and there felt like there was no light in Bri's life.

It wasn't just the fact that she was in constant agony, with other terrifying sensations surging through her body for twenty-four hours a day. Nor that she could no longer do any of the things that brought her joy—raising her kids, laughing with Brian, teaching in her school, helping her community, climbing, skiing, dancing, singing, *anything and everything that made Bri who she was*. It wasn't just because she was no longer a functioning member of her family, and instead felt like such a burden to it. It wasn't that they were rapidly using up their savings and borrowing money from her father *all because of* her, making her acutely aware that she was no longer contributing anything to her family or her community, financially or otherwise.

These all caused Bri pain, but it was nothing compared to trying to get her head around the understanding that the people she'd expected to care in a situation like this—the doctors, the trial staff, the pharmaceutical company, the government—didn't seem to care at all. Brian had reached out to their senator, the House of Representatives, the White House, the FDA, CDC, as well as their local health department. Nothing. *I thought we were all in this together?*

The reality of how she was being treated was beginning to dawn on Bri. She could no longer tell herself that AstraZeneca were too busy; that they'd be in touch soon; that *of course* they cared about her, one of their clinical trial participants. She'd been abandoned. And deep down, she knew it. To them, she meant nothing. Her pain was irrelevant. *Is this what happens in clinical trials? Has this happened to others?*

Publicly, the clinical trials were being celebrated as hugely successful. AstraZeneca was being rolled out in the UK, with no mention of it not being authorized in the United States. And back home, speculation over just how miraculous the other vaccines were was all over every magazine, every newspaper, and every TV show. Social media was awash with campaigns urging America to get vaccinated—*not only do the vaccines prevent Covid, but they cure Long Covid too!*

Doctors, pharmaceutical companies, and governments around the world invested their time, energy, and money, on urging the public to do their part in getting the world back to normal. Getting vaccinated turned into a party offering free donuts and sodas, and, in some states, free joints. The CDC issued cards so that people could prove their vaccination status before gaining access to services and situations that were restricted to anyone that had not been vaccinated. The queues for the cards went around blocks of buildings, with people waiting hours for their ticket to freedom at last.

Bri watched her friends' lives open up as hers became smaller and smaller.

The public were offering up their arms and changing their Facebook profile pictures to announce their vaccination status, assuming that if something went wrong, those same doctors, pharmaceutical companies, and governments would be there to help.

The Dressens knew otherwise.

They were paying a hefty price for the belief and trust they'd previously had. The savings had gone, Bri wasn't earning, and they'd hired an occupational therapist and a physical therapist in an effort to help Bri walk again. AstraZeneca were still nowhere to be seen, and the Dressens had stopped updating the trial clinic on Bri's symptoms and instead just sent them expense updates—the clinic clearly didn't care about what was happening to Bri's body. Maybe they cared about what was happening to her bank account.

Brian continued emailing the clinic proof of the medical expenses and asking them when they'd be reimbursed, as per the consent form. Occasionally, the trial clinic would respond with an apology for things taking so long, or by saying that AstraZeneca would be in touch any day now, or that they hoped to hear something themselves soon. But mostly they didn't respond at all. And nothing happened. Nothing was reimbursed. AstraZeneca didn't get in touch. And the medical expenses kept mounting up.

What happened to all the promises made in the consent form? What happened to AstraZeneca's insurance policy to cover injuries? What happened to the pharmaceutical company agreeing to pay the costs of injuries, including medical expenses? The Dressens went back over the consent form Bri had signed, double checking that their understanding was correct. It was correct—the consent form clearly stated that AstraZeneca would take care of any associated medical expenses; something that the trial clinic had confirmed when they'd been urging Brian to seek out whatever testing and treatment was necessary. *How can they get away with this?*

Bri felt utterly broken. Questioning everything in which she had had so much faith for her whole life—science, medicine, any sense of humanity—was overwhelming and exhausting on top of everything else she was dealing with. Discovering such a lack of ethics within the institutions that she'd so believed existed to protect everyone took Bri to the center of the darkness. Was there anything left to believe in anymore?

God. Can I believe in Him again?

What did she have to lose? Her life couldn't get much worse. She remembered her cousin who had visited and left the letter behind. Her cousin had been so incredibly helpful and supportive during the past few months—spending hours and hours with Hannah, and even more hours emailing everyone she could think of that might have been able to help.

She was a kind, caring cousin, and her letter had been written with love and good intentions. Trish had said that Bri was too fragile to read it, but maybe now was the right time? Maybe God *was* the only person who could help her. She started to pray.

As she prayed, Bri tried to connect with everything good she knew existed within the faith that had played such an important role in her upbringing. Her questioning of Mormonism had never meant doubting the *good* she saw every single day among the people who believed in it— she saw that goodness and kindness now even more than ever as the community had rallied round with everything the family needed during the past few months. And she'd continued to contribute to that community in recent years before she'd become ill, despite her evolving beliefs.

She focused on the God she had believed in during her childhood and prayed to Him—the loving, kind God that she could talk to like a friend whenever she'd needed to. She knew to thank him for the good things and to ask for help with the bad things. She begged for help. She prayed during every waking hour; all day and all night, waiting for the peace and relief that her cousin had promised that only God could bring. Multiple times a day she opened up her heart, bearing her soul to expose all of her pain, in an effort to leave it all at the foot of God. It was true, God had helped her before. She had overcome other things in the past, thanks to His help. Could He do it again?

But praying felt different this time. Every time Bri asked for help it felt like she was reliving everything she needed help with. Her symptoms became worse, and all the physical pain and emotional torment she had worked so hard to control with her breathing flooded over her, instead of the peace and relief she was so desperately hoping for.

Bri became consumed with even more pain and misery. She had begged for spiritual help as well as medical help. Nothing. She had asked for help from everyone—the doctors, the drug company, the only God she knew—and help wasn't coming. She had been looking for a sign to keep going, and no sign had come.

Bri saw the absence of a sign as a sign in itself. A sign to give up.

* * *

Trish could see that her little sister was venturing into a dark place, and in her very matter-of-fact way said that Bri needed to go to the mountains on an adventure with Brian, just like they used to. Bri told her that she couldn't even sit in the car for that long. It was out of the question.

"Nonsense. I'll take you."

Bri didn't have the energy to argue as Trish decided to wrap her up in warm, winter clothes. Just like when Bri was little, Trish picked out Bri's clothing and dressed her. Trish had become an incredibly beautiful woman both inside and out, and had always a great sense of style, even when she herself was young. Being seven years older, teenage Trish had treated Bri like her very own living doll, choosing all of her clothes and dressing her up every day, so little Bri had always looked cute. It had been many years since Trish had had to dress her little sister.

She gently walked Bri to the car. The two sisters had often gone off on winter hikes together, picking wildflowers as they went along, despite knowing it wasn't allowed in America because, well, that's what their mom said they do in Sweden. Bri had kept all of the wildflowers and leaves she had picked over the years with her big sister, and had them stored in an old, thick phone book in the basement, planning on making them into bookmarks for Trish one day. There were a lot of things that Bri had promised herself she would do, one day.

Trish drove them to a spot in the hills an hour away, while Bri sat in the car wearing her sunglasses and earplugs and trying not to think about the pains and shocks going through her body. She switched her focus on to where Trish was taking her and tried to ignore where her mind was taking her. It wasn't working.

"Please, start talking. Talk to me about normal things."

Bri was desperate to get out of her own head. So Trish told her about what her kids were doing, about her work, about her husband, and about all kinds of silly things that the old Bri would have found hilarious. Bri managed a weak smile but felt nothing.

They eventually arrived at their old favorite spot, and Bri tentatively got out of the car, waiting for Trish to attach the ice grips on the bottom of both of their boots before carefully taking one step at a time, telling herself to look around and enjoy the beauty and the silence. Except it wasn't silent because all she could hear was the screaming in her ears. *Breathe.*

On the way to the frozen waterfall, Trish smiled and laughed and did everything she could to try to take Bri out of whatever darkness she was in. Trying to bring her closer to the light.

Bri sat on a rock in front of the waterfall. It *was* beautiful, but she was too physically and emotionally exhausted to appreciate it. Trish realized that her plan wasn't working. She looked Bri straight in the eye.

"Promise me that you're not going to kill yourself."

Bri looked back at the big sister she so adored. The sister who would do anything for her. The sister who Bri *knew* loved her, even if she felt so terribly unlovable. The sister who had taught her how to write, how to tie her shoelaces, and let her bounce up and down on their parents' bed in a tutu so that it would puff out around her. She couldn't lie to Trish.

"I can't promise you that."

Bri knew that something was going on inside her mind, body, and spirit that was spiraling out of her control.

* * *

It wasn't that she *wanted* to die. But she definitely didn't want to live *like this.* Nobody medical seemed able or interested in making "this" stop—not the trial clinic, the drug company, or the doctors. So Bri's mind had worked out a surefire way to make it all stop. If she was dead she wouldn't feel this awful. All. The. Time. She wouldn't be a burden on her family. It would be a relief to them as well to her. There would be plenty of people who would fill in the gap that her absence would bring—the close-knit community they lived in had already shown that with the multitude of people stepping in with all sorts of things that Bri would have ordinarily organized—meals, dog-walking, cleaning, even the Christmas presents under the tree were all thanks to the generosity of their friends and neighbors. *Everyone will be fine without me. More than fine. They'll be better off without me.*

Once the idea of suicide had popped into Bri's head, she found that her mind kept wandering off to the idea. And the more she thought about it, the more appealing it seemed to be. It really was the logical answer. Something to actually look forward to, even. Because after that, all her pain would be gone. Peace, at last.

Bri spent a lot of time fantasizing about ending the pain, undecided on how to do it. She didn't want her family to find her so it couldn't be anywhere near the house. The lake would be a good spot. It would be frozen over at this time of year, but there's always some cracks, and she could just jump on top of one of them, plunge into the freezing water, and not be found until the spring when the ice melted.

Or in the car. Find the strength to drive into the wilderness then park somewhere beautiful, connect the exhaust to the inside of the car, and set the engine running.

Either way, she mustn't forget to stick a big sign on the car window saying, "AstraZeneca did this," just in case there was ever any doubt. Even in the middle of the darkness, Bri knew it wasn't actually *her* who wanted to die. She didn't know who this Bri was—the one that needed to die. She didn't know where the real Bri had gone.

Not that there *would* be any doubt who or what had done this to Bri— everybody close to Bri had seen her turn from a feisty and fun-loving, conscientious and compassionate, daring and driven woman to a hollow version of everything she had ever been. And it had happened within minutes of that jab.

She knew what she needed to do. But she wouldn't do it yet—she had worked out how to do it so that her family wouldn't find her, but that meant that somebody else would, and even if that person was a stranger, she still didn't like the thought of the pain that discovering her body would cause them. Oh well, she wasn't in a rush. She had time to work out the details.

First, she had to make sure that her kids knew how to lead the kind of life that Bri wished for them; the life toward which she was no longer able to guide them. So she wrote both of them a letter with a roadmap for how to be a good person, hid the letters away for when the time felt right, and made a list of loose ends she needed to tie up first. She was always one for making lists and getting things done. This was just another project.

Brian found the letters. It was one of the few times in their entire marriage that Brian lost his temper. He was absolutely fuming with Bri, and fuming with himself. He had been trained by the military to *know* the signs that someone was suicidal, but somehow had missed realizing that Bri was so, so desperate. She had always been so life-*loving* that surely she couldn't be in *that* dark a place within such a short space of time?

He spoke to Bri's therapist, someone she had been seeing for a couple of months at the suggestion of her concerned friends. The therapist told Brian to make the house safer—remove any medication, lock away any firearms, and not to leave Bri alone for more than five minutes. He coordinated with Bri's mom and Trish, and they took turns in watching her. She went from being a struggling adult to being like a child who needed constant supervision, and Brian had to control everything. *What had happened to their life?*

The therapist had a stern word with Bri—they hadn't had many sessions, but Bri had been a good patient and keen to follow the therapist's guidance. The time for gentle encouragement was over and she told Bri that she *had* to find a way to take back control of her brain, despite what was happening to her body. Because Bri had been silently suicidal for months, her brain had rewired into a pattern of thinking that led her to genuinely believe that the strong person that she used to be had gone forever, and dying was now the only option.

* * *

"You *are* strong; you just don't see it."

An old friend had come to sit with Bri not long after Brian had found the letters, and was determined to help Bri see her new self in a different light. Bri rolled her eyes—she was fed up with people telling her that she was strong. They had no idea what it actually felt like to be in this body. The old Bri *was* strong; this Bri was pathetic. Couldn't do anything. Spent her days lying in bed in the dark doing nothing while life went on for everyone else outside.

Her friend continued, regardless, "Strength takes many forms. Volcanoes are strong! You can see their strength as they create lava that moves and shapes the earth, doing incredible things—their power is very recognizable and visible. But glaciers are strong too. They move and shape the earth from *beneath*, providing vital nutrients to the sea life. They are just as strong, but nobody would notice."

"OK, so I'm a glacier."

The thought stuck with Bri. Could she have a different kind of strength now?

She wrote "I'm a glacier" on the bathroom mirror in a dry erase marker, so that every time she looked in the mirror, she was reminded of the strength that she *might* have underneath everything. A few days later, Hannah came into the bathroom, saw Bri's writing, and asked her what a glacier was, so Bri repeated what her friend had said. The little girl climbed up on the bathroom counter and next to Bri's writing, wrote, "Mommy is a glacier! Hannah and Cooper are volcanoes!"

And in that moment, as her daughter embraced and celebrated the strength that her "new" mom had, Bri thought she might be able to embrace it too. She would find a way to reprogram her thought patterns; find a way to ditch the self-pity and self-loathing; and find a new way to live again, *alongside* whatever was happening to her body, instead of fighting it.

Because if she killed herself, then AstraZeneca would have won.

* * *

Reprogramming her thought patterns wasn't easy. There was something that felt distinctly *chemical* rather than emotional in Bri's suicidal thinking. It wasn't just distress at the situation she was in; Bri felt like her entire soul had somehow been taken over by something she couldn't explain. It felt like she'd been possessed—nothing she'd ever experienced before when dealing with other challenging times in her life.

At first, she made a deal with herself. *I will try this for a month and if I don't feel better, THEN I'll kill myself. Maybe two months. Yes, I can try for two months, and if I'm still like this, I can kill myself then.*

It took consistent effort, every single day, multiple times a day, to reprogram her thinking. Every time she had a suicidal thought, she would say out loud to herself, "STOP!" and then remind herself that she *had* to live with this new illness for her husband, and for her kids. Initially, she was saying "STOP!" multiple times a day, sometimes multiple times an hour; her brain had been *that* focused on the fantasy of dying. But within a week, her brain stopped going to that dark place. Her physical symptoms hadn't changed, but something in her brain had.

She found out about suicidal thinking in terms of addiction—something her brain had somehow *learned* was a thought to go to, in order to find comfort and release. It had become a habit. The same way that some people—consciously or subconsciously—turn to alcohol, drugs, sex, porn, social media, or something else, to escape emotional or physical pain, or maybe even life in general. Bri learned to see her own fantasies about dying as something she had actually become addicted to. She needed to teach her brain to expect other thoughts every time she thought about the buzzing inside her body, the screeching in her ears, the blinding pain in her eyes, or the fact that she was no longer able to participate in life the way she used to. She had to retrain her brain so that it went somewhere else for relief, rather than thoughts of her own death.

Her therapist used techniques that had been successful in treating PTSD, specifically EMDR, a reprocessing method used with trauma survivors. The therapist puts their finger up in front of the patient's nose and asks the patient to follow their finger as they move it from side to side, while simultaneously recounting the traumatic experience, in Bri's case being the vaccination itself, its symptoms, and treatment by the drug company and medical system. There were multiple levels of trauma associated with the vaccine, and Bri had been stuck in that "trauma loop" for months. EMDR helped her break away from that loop.

The antidepressants that she'd been put on at first had made her feel much, much worse, but when she tried a different one, Doxepin, she could feel her brain calm down. In addition to its impact on depression and anxiety, Doxepin had a powerful antihistamine element to it that Bri and her doctors felt was helping with neuroinflammation. Before starting that drug, Bri had been convinced that her brain was permanently damaged. It wasn't. Bit by bit, Bri started actually *feeling* a connection again with the world around her. And it was OK that it was a small world.

She could feel a connection with Brian again, who was doing his best to keep the kids in school and food on the table. When the fridge ran bare, the family ate simple meals of just meat and vegetables for a few days. Bri noticed her brain becoming a little clearer, and her confusion a little less intrusive. Brian wondered if their food could be making her worse, and started experimenting, first cutting out wheat and dairy. *Huh . . . she seems a little better today.* Next to go was corn, then soy sauce, and Brian found that each processed food he cut out unlocked a new level of improvement. His wife's personality started returning; there was a little more eye contact and a little less overwhelm. She even reached out to be touched and he wondered, *Is my wife still in there after all?*

Bri became more interested in food again—the months of vomiting had made food a source of distress rather than delight, but her body was starting to accept just a little again. She could keep down very small meals, but it was still terribly painful to eat and would continue to be so for many more months.

Bri made an effort to be kinder to herself, something she'd never really considered in her life before. She had always pushed herself to achieve one goal or another, followed by another one, and then another, treating her body and her mind like a machine. The voices of her siblings had become the voice in her own head over the years. *Keep up, Bri! Don't slow us down. Go faster!* The idea of going slowly and sometimes just stopping to take the time to be gentle with her soul was completely alien to her. But she was determined to embrace this new way of life, so she would lie on the bed with her own hand on her chest rather than Brian's, telling herself that she was loved. And more importantly, that she was loveable, even in the state she was in.

After months of taking responsibility for and intently focusing on her mental health, Bri's physical health became less heavy, less unbearable. Very little had improved in her physical symptoms, but her attitude toward them had dramatically changed. Her brain had learned how to recognize when it was tiptoeing back into the darkness, and instead of responding with, "I need to be dead" she would think, "This is my new life. And it sucks, but I have to live like this, I have to live for my kids."

One day Bri went back to the letters she'd written to Hannah and Cooper, and reread them in disbelief. They were so poorly written with so much of them not making any sense at all—her brain *clearly* wasn't functioning properly at the time. *I almost left them with this. I almost left my kids thinking that a letter would be enough to guide them through life.*

With the brain function now improving, Bri understood that nobody . . . not a neighbor, not a cousin, not her mother, not Trish, not even Brian, and certainly not a letter . . . could possibly replace her in Hannah's and Cooper's lives. Her suicide would have caused immeasurable trauma for her children to carry forever.

She was so relieved that she never went through with it. Yet haunted with a mother's guilt at the thought of what she had almost done to them all.

* * *

While Bri was facing and accepting the reality of her new life, Brian was dealing with the reality of their new financial situation. He kept sending the medical bills over to the trial clinic, and regularly following up to see when payment would be made. The clinic had said that they were forwarding everything on to AstraZeneca and expressed surprise that the Dressens hadn't heard anything from the drug company yet.

AstraZeneca was busy rolling out its product to the rest of the world. While it never gained authorization or approval for use in the United States, it had been approved in other countries, and the first jab was administered to the public in the UK on January 4, 2021. Hungary was next, followed by other European countries after European Medicines Agency (EMA) approval, and by Spring 2021, AstraZeneca was sending millions of Covid vaccines all over the world.

Relieved that the AstraZeneca product had not been cleared for use in the United States, the Dressens were nonetheless worried about what might be happening in other countries that were rolling out the same vaccine that had almost destroyed Bri. *What if the same thing happens to someone else? What if their doctors don't know what to do either? What if they don't have savings? What if they don't have a Brian?* Bri's efforts to reprogram her thoughts may have been working, but she now found herself thinking about others who might end up in the same dark place from which she was slowly emerging. *People will die from this.*

At least there were other vaccines out there. America was getting vaccinated with Pfizer, Moderna, Johnson & Johnson, and there was talk of others in development. The FDA had done the right thing by not approving AstraZeneca: the NIH was taking its role seriously; and its Dr. Nath was interested in looking into what had happened to Bri. If the FDA had authorized all the other vaccines, then the other vaccines must be safe. At least there were safer options available not only in the US, but also around the world.

Brian started looking into the other vaccines. President Biden had made it clear that he expected the entire country to get vaccinated, and Brian had complied, having had two doses of the Moderna jab despite Bri's reaction to hers. He had reasoned that Bri's situation was very, very unusual—there was no need to become skeptical about vaccination in general. Vaccines were an essential part of public health. Where would we be without them?

Curious to see what the science said about each of the different companies' products, Brian found all of the clinical trial reports published on the EMA website, and carefully studied every single one.

When he finished reading the last report, Brian walked into the bathroom . . . and threw up.

"Bri, they're all going to do the same thing."

The Dressens sat in silence together for a very long time, feeling the enormity of what Brian had discovered. This was not good. This was not going to be good at all. There were problems with all the clinical trials, and nobody was talking about it. Other people would be dealing with the same nightmare that the Dressens were.

Bri was not alone.

CHAPTER 5

THE OTHERS

Bri had joined online support groups for clinical trial participants on the day she'd got the shot—you had to show proof that you were a trial participant before your membership was approved. In the groups she had found plenty of people from the Pfizer and Moderna clinical trials but hardly anyone from the AstraZeneca trials, which had only just been resumed in the United States.

The groups were full of participants talking about breakthrough infections and the practicalities of the trial requirements—what needed to be reported and what didn't—but very little about their side effects, which were just the usual ones anybody would expect from a shot. There were a few mentions of tingling, dizziness, and heart problems, but Bri couldn't find anybody having such a *violent* and terrifying reaction like hers. She assumed that it was just her, which made it even more terrifying.

By the winter of 2020, the groups were dominated by discussions about getting unblinded so they could determine whether they'd received the shot or a placebo, and if they had received the placebo, how soon could they get the actual shot through the rollout to the public. The media was constantly emphasizing just how dangerous Covid was, and how important it was to get vaccinated, so it was all anybody on the support forums seemed to be concerned about. The trial participants believed that if you got the shot then you couldn't catch or spread Covid.

The only support groups where people seemed to be having conversations about such debilitating symptoms like Bri's were in the Long Covid groups, so Bri joined those groups too, but the Long Covid sufferers didn't seem to have the electric shock-type sensations that were constantly surging through her body. She never mentioned on those groups that her symptoms were due to injection rather than infection—the lack of support from the trial clinic and AstraZeneca was making her feel a bit confused about the Covid vaccine. She wasn't exactly sure what she thought about vaccines at all now that she knew more about how clinical trials were run. *Vaccines are good, aren't they? We need them, don't we? If I say anything then I might put people off, and I shouldn't do that, should I? Or are they? Do we? Should I?*

But she was scared, felt so alone, and really needed help. Maybe there was *someone* out there suffering the same way that she was. So she wrote a few words about what had happened to her, took a deep breath, and posted it on one of the clinical trial support groups, hoping that it might resonate with someone. Instead, within minutes somebody had taken a screenshot of Bri's post and messaged Bri saying that she was selfish for not going public about it, and threatening to out her, saying the world needs to know what was happening with these vaccines.

Bri was already overwhelmed, trying to find help for her physical symptoms that were keeping her unable to eat, sleep, or think clearly. She didn't have anywhere near the mental capacity to be outed—only a few of her close friends and family members knew about her reaction. Everybody else was under the impression that she had got some weird illness out of the blue. Bri was so distressed that she took her post down and decided not to talk publicly about her experience again—she didn't feel that there was a safe space online for her to do so. She just wanted to get well. She didn't understand why her health was anyone else's business.

But Bri kept looking out for someone . . . anyone . . . who was suffering like she was, as she quietly stayed on the support groups, occasionally commenting about symptoms that she related to. And then she saw someone posting who was also in the AstraZeneca trials! And had had the exact same reaction that Bri had! Sarah Lee, who lived in Georgia.

Bri and Sarah took their conversation off the group and started private messaging each other, with every symptom one of them mentioned being met with a "Me too!" from the other. The similarity of their experiences was uncanny, right down to the timing of when they both got a kidney infection. They had had the same response from their doctors, from their trial clinics, and from AstraZeneca. And they had the same conflicting feelings about what was happening to them, and were taking the same approach by keeping everything very, very quiet, and dealing with everything alone.

Except now they weren't alone.

Bri was so excited. Now she had someone to talk to! Someone who understood *exactly* what she was going through! She was elated and relieved, but then . . . she became even more conflicted at the realization that it wasn't just her. *Surely if two of us had the same reaction they would have shut down the trial?* The consent form . . . the trial clinic lead investigator's attention to detail . . . the fact that they paused the trial before . . . all of these things showed that AstraZeneca took these adverse reactions seriously. Didn't they?

Brian then started pushing the NIH to move forward with their study, and Dr. Nath took up Sarah's case. *Now there are two of us! Dr. Nath will fix us!*

Within days Bri then met Olivia Tesinar, another clinical trial participant, and Bri and Olivia had a similar conversation via private messaging. But Olivia hadn't been in the AstraZeneca trial. Olivia was in the Moderna trial. *Brian was right! It isn't just AstraZeneca!*

Unlike Bri and Sarah, Olivia was being the exact *opposite* of quiet. She was determined to tell anyone and everyone about what was happening to her, and her Facebook page was full of photos showing the reality of her reaction. Olivia's vaccine reaction left her with lymph nodes so swollen and deformed they had to be surgically removed, leaving her in a massive amount of pain. There had then been complications following the surgery, so she was now bedridden, and covered in bandages with tubes coming out of her to drain what the lymph nodes hadn't been able to cope with. Moderna weren't paying for any of it or helping in any other way. Olivia wanted to make sure nobody else fell to the same fate and didn't care who knew about what happened to her.

Bri kept an eye out for any other hints of adverse reactions in the trial participant support groups, confused that Moderna had been authorized despite the symptoms that Olivia had reported. *Surely Olivia's reaction warranted an investigation?*

On the Long Covid groups, everyone was talking openly about their symptoms, and as the rollout progressed, people started asking how others were getting on with their Covid jabs. Bri nervously mentioned that she had a problem with hers, and a little comment popped up saying "me too." *Ooooh! Another one!*

Dr. Danice Hertz was a recently retired gastroenterologist who had been very worried about catching Covid and was thrilled to get the Pfizer jab early in the rollout. Her adverse reaction had been rooted in her mast cells—cells that release chemicals to help the body deal with allergies. MCAS (Mast Cell Activation Syndrome) is when the body releases so many of these chemicals that they can't be processed quickly enough, so Danice was suffering with ongoing allergic-type reactions to everything around her—foods, drinks, fragrances, household products, even stress and pain—and she never knew when the next trigger was going to hit. It was completely debilitating. Like Bri, Danice experienced debilitating pins and needles and internal vibrations, but also experienced what felt like a tight band around her chest, tinnitus, and severe nerve damage to the face. Her symptoms also kept throwing themselves at her one after another, with no warning or explanation, and also like Bri, she had nowhere to turn other than to find others like her.

Danice was resourceful, determined, and very well-connected in Southern California. She had been dismayed to discover that her colleagues and contacts had no idea what to do, so had been in regular contact with the FDA since the end of December 2020, expecting the government to be interested in what was happening to her. The FDA had been responsive and confirmed that they were getting reports of reactions that were similar to hers, but they didn't have any pathways for treatment yet. *More people in the government know and care about us! They're going to help!*

Bri arranged a meeting between Danice and Dr. Nath, and Danice didn't just join the study, she brought more and more injured patients with her, and the NIH added them to their study too. Danice was on a mission to find more vaccine-injured people out there, always reading scientific websites and medical journals, and if there was an opportunity to comment, she'd describe herself as a "vaccine-injured physician." Soon she was inundated with emails from others who were the same.

Sheryl Reuttgers had been studying to become a psychotherapist in Oregon when she'd had a severe neurological reaction to one shot of the Moderna vaccine. She hit it off with Bri immediately. They not only shared the same symptoms, but also the same sense of humor, and pretty soon, Bri was able to laugh at some of the ridiculousness of her terrible situation. *I just threw up all over my kid, now he knows what it feels like LOL.* Or as the weight dropped off them both, *all you need is a vaccine injury and you're ready to be a Victoria's Secret model!*

Sheryl was married to Ken, who was a former Super Bowl champ and Green Bay Packers Hall of Famer. He was hugely inspirational to other players and a famous champion in the nineties, but Bri had no idea who he was. All she knew was that his wife Sheryl was an inspiration with her ability to help others experiencing hardships, her fantastic sense of humor, and her infectious laugh, despite all the challenges she was dealing with. The Reuttgers and the Dressens had much in common, with both couples coming from Christian backgrounds, and both raising their children in similar ways. They would spend a lot of time on group calls discussing the latest papers that Brian had read and speculating on what could be going on with the vaccines, talking for hours about the spike protein.

Mary Johnson joined the small group of injured women. Mary was an ER and critical care physician and had played a significant role in her local community's management of Covid. She was the only person in her hospital who could do intubating, so had been very much in demand throughout 2020. She had been on the front lines. She had seen the very worst of Covid and had just begun to see the worst of the vaccines as well when she became severely injured herself, driving her out of her position in the hospital and confining her to her home. It had been a dramatic interruption of life for this young newlywed with a successful career.

Given how much Mary—and other healthcare workers—had been such a big part of the Covid response under such incredibly difficult circumstances, it was shocking for Bri to see just how little help there actually was for people like Mary after a vaccine reaction. Mary had had her entire life ahead of her, and now, couldn't even get off the couch. Dreams of a family, the career that she trained a decade for, a full life . . . all gone.

Danice found two more—Kristi Dobbs and Candace Hayden. Kristi was a kind and compassionate dental hygienist in Missouri, who was also a supermom of four young kids. Her family was very dependent on her, and the Pfizer vaccine had completely disabled her with symptoms that were very similar to Bri's. Michigan-based Candace had been a regular triathlete prior to her Moderna jabs, but had developed transverse myelitis after her second, so could no longer even walk. She was determined to fight her way through this, and would be using her walker around the room when she joined group calls, despite having been told that she'd never walk again. Candace was tough. She was also very smart, with a good instinct for business, and an even better instinct for bullshit. Candace was very suspicious of the NIH.

Bri completely trusted the NIH to take care of them all. Dr. Nath had given her the impression that he and his colleagues were going way beyond their job descriptions and the NIH responsibilities in paying such close attention to them all. Bri was honored that they were all getting such special treatment. Dr. Nath was so kind and would make so much time for them, even making the effort to talk to them on Sundays on his cell phone. He was soft-spoken, and his voice felt like a rare source of comfort coming from someone who represented the government. He always signed his emails "Avi" rather than using his full name or title. The injured group nicknamed him Yoda.

As more and more injured people were found, Danice and Bri realized that they couldn't just refer everyone over to Dr. Nath and his team. There were just too many injured and not enough staff at the NIH. So they introduced only the sickest or the most desperate—it was a heavy responsibility to bear, but they carried it while holding on to the belief that the NIH's support of a small number of injured would lead to support for them all.

The NIH team validated all their injuries and were happy to speak directly to individual doctors and specialists on behalf of the study participants. They confirmed that they were receiving reports of vaccine injury at the NIH, encouraged doctors to take their patients' concerns seriously, and suggested different treatments that they could try with these complicated patients. Bri was relieved to at last have *somebody* official not only listening to her—and her new friends—but also taking action on their behalf.

And they were relieved to have found each other. Bri, Danice, Sheryl, Mary, Kristi, and Candace set up their own group chat not just to talk about how the NIH study was going, but also to share information about what was helping them. It was such a relief to not have to spend half an hour explaining everything to someone before answering that simple question, "How are you?" They just *got* it.

Bri was in awe of this group of accomplished women she was spending more and more time with online. They reminded her of Trish—straight to the point, no messing around, and really clever. They were all highly educated, comfortable with medical terminology, and politically aware. Brian would sometimes join their chats and share the latest research he'd come across, and Bri was happy sitting quietly in the background, willing to learn from everyone, and willing to follow Danice's lead.

Danice had connections in the medical world that the rest of them didn't—healthcare workers were among the first to get vaccinated so naturally were among the first to get injured. Danice was also making some headway with other researchers, and within weeks there had been scores of them connected with the NIH. Bri couldn't believe how many people with adverse reactions were out there, and how many of them were being ignored by the doctors and the drug companies. *Wow! Hundreds of people suffering like me! How many do they need before it gets pulled?*

The five women kept reassuring new study participants—just as desperate as they had all been—that help was coming. The NIH study was going to find answers. The NIH was flying people out for assessments; everyone just had to be patient and wait until it was their turn. The NIH *would* take action. They would make sure that the trial clinics, the vaccine manufacturers, and the medical professionals paid attention to this. The NIH would bring about change for anyone that was struggling as a result of the Covid vaccine.

Danice and Kristi decided to make a Covid vaccine injury support group on Facebook. The social media platform had served the Long Covid groups and clinical trial participant groups well.

Other vaccine injury support groups started popping up on Facebook too. Danice scrolled through them looking for other injured medical professionals to invite to their group, while Bri focused on reaching out to anyone and everyone that had been impacted. Candace joined the very clearly titled "Covid Vaccine Support Group" set up by social worker Angela Hammond in Florida. Angela's group had been going for a month and it had *thousands* of members, some of whom weren't injured themselves but had some very strong opinions about the vaccine and those that it had harmed. *People know about us!*

The dissenting voices were a shock to Bri. The media that she and Brian followed had never featured any message other than "get vaccinated, it's safe and effective" and she had no idea that there were people outside of the "Disinformation Dozen" who were not just questioning that message but actively promoting quite a different one.

The Disinformation Dozen list had been released in March 2021. It contained the names of twelve healthcare professionals, authors, and activists that were considered to be the biggest vaccine-related disinformation spreaders by the British non-profit organization, the CCDH (Center for Countering Digital Hate). The list had been widely publicized in the United States, with even the White House emphasizing the danger of this group of twelve and the questions that they were asking.

Bri put her own questions to one side—something she was finding harder and harder to do as she was learning just how many injured were out there—and being desperate for answers herself, she concentrated on connecting with the injured instead. People were fascinated by her experience and incredulous at how she'd been treated as a clinical trial participant. She was welcomed everywhere. She found another group of thousands; these were people who had started suffering with tinnitus after their vaccine. She found a smaller group of people that were dealing with tingling, and was asked to become an administrator there. Then she found other groups focusing on other symptoms, and soon she realized that there were thousands and thousands of people out there, all talking about adverse reactions *they themselves had personally experienced.*

There most certainly *were* others. And they were everywhere. And they all had the same story, they were all struggling to get support, and they were all spending every cent they had, with very, very little to show for it.

The Dressens were still sending all their medical bills over to the trial clinic. If the clinic replied, it was always to assure Brian that they were still forwarding everything to AstraZeneca, who would be in touch any day now. Bri had emailed the clinic in frustration at one point, asking for the contact information for whoever would be party to legal proceedings, and the clinic had responded, confirming that they would forward AstraZeneca's legal contact to the Dressens the following day. *Whoa. They're not going to cover this. That consent form meant nothing.*

Reality kicked in. The Dressens had been strung along *for months*. The costs of Bri's injury (which AstraZeneca had promised to cover) continued to mount. They had lost Bri's income, used all their savings, borrowed tens of thousands of dollars from everybody close to them, and had ongoing medical expenses for a condition that nobody understood or talked about, not to mention the addition of childcare costs.

Hannah and Cooper had by now become so distressed by their mother's ongoing ill health that they were really struggling in school. Cooper, in particular, had found it very difficult to even leave the house, afraid that when he got home, his mom might not be there. He hated it when his parents would have to go to the hospital, and waited anxiously for their return, upon which Cooper would run out to the car to check that his mother was actually in it. Bri may have been making huge steps in accepting the reality of her new life, but the children were struggling.

Over 1,600 miles away in Cincinnati, another child was struggling. Twelve-year-old Maddie de Garay.

Maddie was a vivacious young lady. She was full of life, always playing sports or having fun with her friends. She had a great sense of humor. Until she got a Pfizer vaccine.

Maddie's mother, Steph, had enrolled all three of her kids in a clinical trial. Maddie had reacted immediately, and by the time Candace found Steph, her daughter had been in hospital three times for a total period of more than two months. Maddie was unable to walk, eat, or go to the bathroom, and had all sorts of tubes attached to her—a G-tube (Gastrostomy tube) inserted through her nose to deliver nutrients to her stomach, and a catheter to empty her bladder. She suffered from severe seizures, fainting episodes, and also had the same ongoing internal vibrations that most of the other injured had. She was in excruciating pain.

The doctors said it was all in her head.

Steph had been advised to take away Maddie's wheelchair because then she'd *have* to walk, so following the experts' advice, Maddie was dragging herself across the floor, trying her best to be a good patient, and moving her legs back and forth and doing whatever else the doctors told her to. It was heartbreaking, and Steph had no idea what to do. The trial clinic wasn't helping. Pfizer wasn't helping. And the doctors weren't helping; they refused to even *explore* the possibility that a vaccine could have done this, and Steph was afraid of saying anything that might lead to a withdrawal in her daughter's care. Steph was terrified, desperate, and felt so utterly alone.

Bri reached out to Steph, saying that she'd been in a clinical trial too, and telling her that the NIH was researching everybody. Steph was so relieved. *Somebody* understood, and somebody was going to help. Bri suggested that she connect with Dr. Nath and the NIH. Of course, Dr. Nath will help turn things around for Maddie; after all, he was helping so many others by now.

Steph spent days preparing all the information, feeling like this could be Maddie's only chance at getting help. She emailed everything to Dr. Nath, who responded by saying that he would be happy to speak to Maddie's doctors. Steph cried and cried, so relieved that Maddie's case would be getting guidance from none other than the lead research institution in the country. She and her husband had been in survival mode for months, trying to keep a roof over their heads, trying to keep themselves together for the sake of their other kids, and trying to keep their daughter alive. At last, someone was going to help. If the NIH tells Maddie's doctors to help her, they will help her.

Finding Maddie had an enormous impact on Bri and the core group of other injured who were becoming active in supporting others—Danice,

Sheryl, Mary, Kristi, and Candace—all of whom were mothers themselves. It was one thing to be dealing with your own adverse reaction and the multiple levels of disability, distress, and discard that involved. It was another thing entirely to watch a twelve-year-old girl *who participated in a clinical trial* go through it.

It was time to go public.

PART TWO

THE INJURED

CHAPTER 6

GOING PUBLIC

Bri and her new friends thought that any reasonable human being—regardless of their opinion on vaccines—would have been *appalled* at how they were all being left behind in the wake of the rollout. Anybody with a heart couldn't fail to be moved by Maddie's situation—the thought of a twelve-year-old clinical trial participant dragging herself along the floor of a hospital while a medical team refused her wheelchair access haunted Bri. *The public would care about this, surely? If only they knew.*

Sheryl Ruettgers' husband, Ken, not only had media experience through his football past, but also now had family ties in marketing, and volunteered to make the injured group a website. Bri worked closely with Ken on putting together some basic information—who they were and what they were trying to do, and encouraged other injured to send in their stories.

Bri's job was to go through all of the stories that were submitted. She read every single one before posting the testimonies on the website. The stories were harrowing; so many people suffering in the exact same way without any support. Bri saw herself in each and every one of them; each experience left an imprint on her heart. Her desire to help grew stronger, along with a growing sense of urgency for the NIH to figure this all out so that relief could be provided for so many people who were clearly suffering.

In some aspects, working on the website helped Bri with her own health—there was the reassurance that she was not alone, and a sense of purpose to her days in knowing that she was playing her part to support others in this new community she had found herself member of. But reading about all of the pain and suffering took its toll on Bri too. Nevertheless, she persisted, knowing that there were more people out there who, like she had, would be thinking that they were alone in their suffering.

Ken suggested that the website needed to be more interactive, so he created video testimonials of Bri, Danice, Kristi, Candace, and Sheryl. They talked about their experiences after the vaccine, and the videos were added to the website along with the stories. The videos found their way to TikTok, and suddenly they went viral, with a million views . . . two million, three million . . . the numbers just kept going up and up, the traffic to the website went crazy, and suddenly Bri was inundated with story after story about somebody's vaccine injury. And her own video seemed to be getting way more views than the others, which was a bit scary at first—she didn't want people looking her up on Facebook and seeing pictures of her kids. But it was done, she was out there, and everybody knew who she was.

Bri had no desire to be public at all—none of them did. Danice was extremely uncomfortable about it. But it was something they felt they had no choice in—it was necessary to get the support that was needed for themselves and the increasing number of other injured that they'd been meeting in the online support groups. Bri had even changed her surname on her Facebook profile, hoping that that would be enough to keep her anonymity, but the name change only resulted in friends asking her if she and Brian were splitting up.

They still hadn't told most of their friends and family that Bri's mysterious illness was anything related to a vaccine. Despite the past six months and despite the unsettling questions about vaccines that had started to pop into their minds, the Dressens didn't want to be responsible for people questioning whether they should take one or not. *Vaccines are good, aren't they? We need them, don't we? Volunteering for the clinical trial was the right thing to do, wasn't it?*

One of Bri's brothers didn't even know that she'd signed up for a clinical trial, and she was dreading telling him. He had very different political views to the Dressens, and had been questioning everything about

Covid—social distancing, masks, and the jabs. She hadn't had the energy to have an "I told you so" conversation with him before, but now that she was public, she felt that she had to.

She said words she thought she'd never say to her big brother, with whom she had never seen eye to eye politically: "You were right. You were right about all of it."

As they became public about their injuries, for much of the injured community, this would bring about unexpected challenges in their relationships, as some friends and relatives distanced themselves due to all the discomfort involved around the topic of vaccine-related harm. Bri's siblings were all supportive, but not all of her friends were, including the two friends who'd participated in the Moderna trials; they chose to quietly move out of Bri's life after her adverse reaction.

However, as Bri became more public, there were plenty of others moving *into* her life, one of whom was Heidi Ferrer.

Heidi was a very successful screenwriter in Southern California, who had worked on *Dawson's Creek* among other projects, but had been struck down by Long Covid. She managed to get back to 75 percent of her previous functioning, and regularly blogged about everything she was doing to help herself, keenly anticipating the vaccine which the media was promoting as a treatment for Long Covid. But when Heidi had gotten the vaccine, she had become far sicker than she ever was with Long Covid, and her functioning decreased to only 15 percent of her pre-Covid self. Her symptoms were mainly neurological with severe pain, hands that constantly shook, and feet so stricken with neuropathy they couldn't bear her weight. Having spent months pinning all her hopes on the heavily publicized promise that the vaccine was the way to repair health from Long Covid, Heidi had been couch and bedbound by the vaccine itself, and was getting desperate.

* * *

The website was getting over 300,000 views a week, stories were pouring in from all over the world, and videos were being shared. More and more support groups were popping up all over social media, but especially on Facebook; some groups were specific to side effects, some specific to countries, and some specific to the brands—Bri found a Moderna group, a Pfizer group, and an AstraZeneca one. AstraZeneca hadn't been administered in the United States so until then Bri hadn't found many fellow AstraZeneca-injected, but by the spring of 2021, it was what the majority of Brits had been given. Reports of blood clots were circulating in Europe, and the brand's Covid vaccine was ultimately limited to those over forty years old only, but nonetheless, AstraZeneca had been celebrated as a tremendous success in the UK media.

Kristi, Sheryl, Danice, and Bri started reaching out to the US media, knowing it would be a difficult task. Practically as soon as Covid happened, the mainstream American media had been touting the Covid vaccine as the key to saving the country, and the world. There had been countless stories about how safe and effective the vaccines were, and about how it was *vital to society* that everybody got vaccinated as soon as possible. The injured knew what they were up against but persevered anyway, contacting hundreds of local and national media outlets. *If the media was all so pro-vaccine, then they would definitely want to know about what's happening to us and play their part in helping to make vaccines safer.*

Danice landed an interview with a Pittsburgh reporter, and she set up a group call with the others. The call went on for hours, as each of the injured shared their stories with the sympathetic journalist who listened, incredulous, occasionally saying, "Wow. Just wow." They all cried during the call, including the journalist, who said that she would get statements from the government and the drug companies, and promised to run the story soon.

Feeling encouraged, they continued reaching out to other journalists in their local areas, and Bri managed to secure an interview with a reporter in Utah, who also promised to get statements from the government and the drug companies, and that she would run the story soon.

And then some of the bigger media outlets responded—the *LA Times*, the *New York Times*—at last!!! We're going to be heard!

The original injured group had been expecting that any day now Dr. Nath would have all the answers and treatments that everybody needed. Bri was so confident that she introduced several sympathetic news reporters to

Dr. Nath so they could get some quotes directly from the NIH. It was only a matter of time before the NIH study would be finished, this would all be in the mainstream media, the government could arrange for treatment pathways for everybody affected, and they could all just get back to their own healing and what was left of their lives.

They knew they had NIH support, but in the meantime, they started planning on how they might reach out to other government agencies in a more formal manner. Many of the injured—and not just Danice, Bri, and co., but also many of the online support group members—had tried contacting the FDA, CDC, and VAERS, but *nobody* appeared to have received any significant response. Perhaps if they reached out to the government bodies more formally, as an entire group and not as individuals, then they might get a response. They needed a very well-written petition, signed by as many of them as possible, and sent to the top health officials responsible for—and publicly claiming to be—addressing vaccine harms.

Brian drafted a very science-based letter that could form the petition, and Danice edited it. Bri added one tagline—"PLEASE HELP US"—and started organizing signatures along with full names, exact locations, and vaccination batch numbers. The petition needed to be as clear, detailed, and professional as possible. It took an *extraordinary* amount of time and focus, not just in paying attention to all the details required, but in responding to everyone who requested to sign the petition. Bri read every message, and replied to every single one, often with tears pouring down her face as people would frequently send her long emails describing everything that had happened to them, how nobody was helping them, how terribly afraid they were, and how alone they felt. Bri's inbox was flooded with extremely unwell and *very* scared people—"I'm so desperate for some relief." "Has been six months now with no end in sight." "I feel as though my time is running out."

They were the people that Brian had *known* would be out there when he'd read all the clinical trial data months earlier. He was right. *Why wasn't the government helping all these people?*

People weren't just finding Bri's contact information online, they were also finding her on Facebook, despite her attempts to remain elusive. After her videos had gone viral, she had been inundated with friend requests from people all over the world. While she was working on the petition, there were too many to even go through to accept or reject the requests, so she just stopped trying to process them all and stayed focused on getting the petition across the finish line. Time *was* running out.

And on May 26, 2021, time ran out for Heidi.

Heidi hadn't been a stranger to complex illness. She had been her only son's champion as he struggled with spina bifida and was determined to bring joy to her family's life with her bright personality and warmth. Her husband described her as "sunshine in a dress." Bri had detected Heidi's loving nature in every one of their conversations; the two of them would often message through the night while surrounded by darkness.

Hey, having a hard time. Can you talk? Heidi x

Bri would talk, urging Heidi to hold on. The NIH were working on it, some people were finding some relief after trying a few dietary changes, this medication or that, maybe a new supplement. Bri would tell Heidi that she was slowly getting better too, even though she wasn't really. She just wanted to give her new friend a little bit of hope; a reason to carry on.

And Bri did give Heidi that hope. Heidi would end the message with relief and a smiling emoji, telling Bri how strong she was. And Bri didn't know what to say in response, so she didn't say anything. She didn't tell Heidi that she had been in a very, very dark place, and that she still returned there some days. Bri didn't feel strong at all.

Some time after one of the pep talks Heidi's hope started diminishing to the point where it disappeared completely. Her husband became concerned, then remembered Heidi mentioning a new friend who had been injured in a trial. She always seemed to perk up after speaking to that friend. He sent Bri a friend request, a call for help. Bri didn't see it among the flood of friend requests.

A few days later, Heidi's husband had left her alone for just five minutes, only to return to the sight of his wife hanging from their four-poster bed. He ordered their son to go to his room and stay there until told otherwise, took Heidi down from the beam, and held her in his arms as he tried to revive her. It was too late.

Bri was devastated. *What if I'd responded to his friend request? What if I'd checked in with her more often? What if I'd been public about my own injury earlier? What if, what if, what if? What if I'd told her that I've been there too?*

Bri was still fighting her way out of her own darkness. Every day was still a battle with desperate thinking about how she was no longer the person she used to be. Bri wasn't preoccupied with her own death the way that she had been not that long ago, but she was still in a very fragile state, trying to love herself as a glacier instead of a volcano.

Bri knew exactly why Heidi had ended her life, and she didn't blame her at all. Bri blamed herself.

After a long and tearful conversation with Heidi's husband, Bri was then tasked with informing the rest of the group. The women sat on a call together in silence, as if the air had been sucked out of their lungs all at once, until Danice said, "I know what she felt like," followed by Kristi, "Yes, I was there," and a few "Me too"s, and then Bri built up the courage to sheepishly share that she had also been there. All of them had been—and admitted that sometimes they still were—in a very, very dark place.

Bri was stunned. Here were these incredible women whom she so admired: intelligent, empathetic, feisty, capable, butt-kicking *dynamos* of women who had achieved the most incredible things in life both person-ally and professionally, and this vaccine had brought them to their *knees*. It had broken them in ways that not one of them had admitted since they found each other, despite sharing everything else. *How many others are out there with the same utter exhaustion, the same relentless pain, and the very same overpowering thoughts that we all have?*

They decided that as vaccine-injury support group leadership, they had a responsibility to talk openly about the full impact of being vaccine-injured. About everything. Every ugly aspect of it, including the intru-sive and suicidal thoughts that were perhaps more common than any of them had thought. Whether these thoughts were chemical or emotional in origin, other vaccine-injured needed to know that they were not the only ones having them. Telling people they were not alone didn't mean telling them, "I'm here for you." It meant telling them, "I've had those thoughts too."

If Heidi had known that she was not alone with those thoughts, could her life have been saved?

Shaken, Bri completed the petition, and submitted it to the FDA, CDC, VAERS, and the White House within days of Heidi's death.

To Bri's surprise, Janet Woodcock herself—the FDA's head commissioner—responded. Danice wasn't holding her breath. She had been going back and forth with Woodcock for months, and it was going nowhere.

Janet connected Bri directly to Dr. Peter Marks, Director of the Center for Biologics Evaluation and Research at the FDA. Dr. Marks had been responsible for the licensing of the Covid vaccines, and for ensuring rigorous safety and efficacy standards both before and after licensing. He worked closely with the CDC and had the final say on whether any product reached the public or not. The *Financial Times* had reported that Dr. Marks ". . . personally told AstraZeneca staff that they should complete their US trials before applying for emergency approval" so Bri thought he might be especially interested in her experience of the AstraZeneca clinical trial, which was still ongoing.

Dr. Marks had also been involved in the partnership between the government and pharmaceutical companies to ensure speedy availability of the Covid vaccines. Then he had been involved in the FDA's "just a minute" campaign where he would answer questions and reassure the public of the importance of getting their vaccine.

The same guy who was tasked with monitoring the safety of the shots was also tasked with getting those shots into arms.

Bri and Danice were now talking to the most senior people in the regulation of pharmaceutical products *in the entire country*. There was nobody else they could go to. They'd done it. They'd gotten the top people in the country to speak to them.

Bri updated the NIH to say that they had Dr. Marks on board, and assumed that the different government bodies would now be working together on finding a solution to the life-changing—and in Heidi's case, life-ending—health challenges that the vaccine-injured were facing. By now, word had spread within the vaccine injury community and the NIH had been introduced to *hundreds* of people with adverse reactions to the Covid vaccines; some of whom the NIH were very quietly flying out to their headquarters at a rate of at least two per week.

In June 2021, eight months after she'd been vaccinated, it was Bri's turn to go to the NIH.

Bri hadn't been anywhere since that drive with her big sister. She had smiled for the selfie that Trish had insisted on taking of them both in front of the frozen waterfall, but she'd been in severe physical pain. Trish was terrified that it could be the last time she spent with her little sister, and when she'd returned to work, the waterfall outing had prompted the usually calm and professional Trish to stand in the middle of her colleague's busy office and yell at the top of her voice, "If somebody doesn't do something, my little sister is going to off herself!" Everyone had gone silent, and the head internist rearranged his schedule to see Bri.

Bri had taken to her bed as soon as she'd got home from the waterfall; her ears screaming and her entire body vibrating. Just a little bit of movement meant nightmarish electrical pain. Now she was on her way to the National Institutes of Health headquarters in Maryland, a four-and-a-half-hour flight away.

The NIH headquarters was *not* your normal healthcare facility—in the wake of the January 6 insurrection, the NIH had full security detail complete with bomb sniffing dogs, X-ray machines, and security staff. All of Bri's belongings were carefully checked—even the bear that she took as a gift for Dr. Nath. She'd chosen a bear as a symbol of bravery, acknowledging how brave she thought the NIH was being for putting so much effort into investigating the growing community of vaccine-injured. She was hoping that he might put the bear on his desk or at least in his office somewhere, and not forget about those who had been damaged by the Covid vaccines.

Bri and her new friends had all detected a distinct unease among most of the doctors and nurses they had engaged with during their multiple hospital trips, as well as a general lack of knowledge about what to do in the case of an adverse reaction. Vaccine damage wasn't only something doctors seemed to be unaware of, they didn't seem comfortable talking about it either, and it was hard to pinpoint exactly why.

The Dressens wondered if this reluctance had anything to do with the Disinformation Dozen list. *Are doctors afraid of talking about vaccines? Afraid of losing their jobs? Of being put on some list?*

Brian had been surprised to find Robert F. Kennedy Jr. on the list—if someone as well-connected and influential as a Kennedy could find themselves on a very public "list of liars" for questioning vaccines, then Dr. Nath and his colleagues deserved to be appreciated for their bravery in working with the vaccine-injured. *Maybe that's why everything feels a bit hush-hush. The NIH just need to get all their information together before they'll go public. They can't afford for there to be any questions about their work.*

The NIH put the Dressens up in a nearby hotel, and every day for a week they were transported to and from the healthcare facility. Each day brought a different set of tests; some of which had already been conducted during one of the many hospital visits during the preceding months but had not resulted in any treatment. The tests confirmed the physical damage that had been done throughout Bri's body; nerve damage in the leg that was slumped the morning after the shot, memory loss, and issues with Bri's autonomic nervous system that explained the wide range of problems affecting her digestive, urinary, circulatory, and cardiovascular systems. It most definitely was *not* just in her head.

Bri had spent the week having her entire experience validated by the NIH team, who diagnosed her with "post vaccine neuropathy." It was agreed that IVIG—the treatment that Brian had begged all the hospitals to try—was the best course of action, bringing about some very mixed feelings. At last Bri had some diagnoses and possible treatment pathways, but what damage might have been done in the meantime? *Am I ever going to get better?*

The NIH doctors reassured Bri that she *would* get better, but it would take time, and she had to be patient. So much was still unknown about these new vaccines. They did, however, know that the vaccines weren't preventing or treating Long Covid, despite what the media was saying. More research was needed.

Bri mentioned some other researchers that Danice had introduced to their little group—the Mayo Clinic was putting together a series of case studies to be published in a medical journal. Dr. Nath frowned, saying that it wouldn't get published. Case studies weren't *actual* research. Best to leave it to the NIH to produce reports on vaccine adverse reactions. They were working their hardest, but it was probably a good idea to keep quiet about the study for now, as Bri was reminded when she was discharged, "We know you and your friends are talking online, but we need you to be quiet until we know more and get this publicized appropriately." Bri had agreed—they were arranging treatment that could dramatically improve her quality of life after all. *But how do they know my friends and I are talking online?*

One of Bri's new friends was an NIH employee.

Casey had suffered a severe neurological reaction to her Covid vaccine and was a member of one of the online support groups, which is where she heard about the NIH study. Simultaneously curious about what her employers were doing, and annoyed that they hadn't asked her to participate despite her begging them for help for months, Casey had looked into the study and decided that something didn't seem right to her. Casey had aired her concerns directly with Bri the night before Bri had flown out to the NIH.

Casey knew how NIH studies were set up, and she could see that the vaccine injury one was set up differently. She decided that the NIH's Internal Review Board needed to know about it. The IRB protected study participants by ensuring that studies followed certain ethical guidelines and ensured the integrity of data. Casey assured Bri that her actions wouldn't impact Bri's access to support from the NIH, but it might affect others'. Bri was shocked by Casey's revelation; less concerned about her own need for help and more concerned about the hundreds of people she now knew to be very, very desperate. *We can't lose any more people. We NEED the NIH study. It's our only hope.*

Bri met with Casey face-to-face inside the NIH and begged her to change her mind, but it was too late—she'd already had a meeting with the IRB during which Casey had confronted Dr. Nath, asking him about his research: why it wasn't following standard NIH protocols and why she was being excluded from it.

Casey said that Dr. Nath looked her straight in the eye and flatly denied that there was *any* research going on.

Casey said that she had kept pushing, and Dr. Nath had continued denying that anything was going on. In desperation to not be dismissed, Casey then named every single person she knew was participating in the study, and every single person that had physically been to NIH headquarters, "Are you really going to keep denying this is going on?"

Is this how they know we're talking? Bri's heart sank. *The NIH might not be able to save us after all.*

Bri flew back to Utah relieved to have some help for herself, but sick with worry about what might happen to others. She told the other advocates what had happened, and they were just as shocked as Bri had been. *Why would the NIH go to all that trouble of flying everyone out there? Speak to our doctors? Arrange our treatment? Then deny it? What reason could they possibly have?* Nobody wanted to believe that the NIH could be conducting an unethical study, but something wasn't adding up. Bri told herself that NIH were doing their best, the FDA knew about the NIH involvement, the media were busy getting quotes from all the government agencies, and the injured needed to keep their focus on their continued awareness campaign, which also included senators.

Many of the injured had individually reached out to their own senators when they had first become sick, and some were regularly following up, with very little response. Any senators who did respond, would say the same thing . . . *report to VAERS . . . never heard of this kind of thing happening . . . very rare . . . safe and effective . . . the best way out of the pandemic . . .* blah blah blah. Nobody responded with anything that could actually be of any use and very few responded with any empathy. Despite the advocates' optimism, their attempts to communicate with senators as a collective simply generated the same heartless and fruitless responses that had been received when they'd communicated as individuals.

At a less than sixty percent take-up of the first dose, and about fifty percent take-up of the second, there was talk of the vaccine being mandated throughout the country. Senators seemed reluctant to be saying or doing anything that could be seen as speaking out against that. They didn't even seem willing to speak privately. The general consensus among senators, and most other influential people, seemed to be to avoid sticking your neck out on the topic of the Covid vaccine in any way, shape, or form.

Nevertheless, Ken Ruettgers had been calling anyone and everyone he knew, using the contacts and credibility he attained during his days and continued connections with the NFL, to see if *anybody* would be willing to have a conversation about what was happening to his wife and the other vaccine-injured. Or even just to listen to them. The only person who responded was Senator Ron Johnson.

Wisconsin's Senator Johnson had recently made headline news because YouTube had suspended him for his comments about the Covid vaccine, which he claimed had resulted in thousands of deaths. He was a highly controversial figure, portrayed in the mainstream media as being an extremist. He had questioned lockdowns and provided platforms for health professionals to discuss ways of managing Covid that were not limited to vaccines, and none of his actions had made him very popular in the socio-political climate of the time.

He was the last person that the vaccine-injured wanted to be associated with. All this time they had been so careful to try *not* to stir up any trouble or make anybody feel uncomfortable; for some reason always feeling the necessity to state a pro-vaccine stance just to be heard. But they weren't being heard by any politicians—their mere existence was controversial enough. They couldn't afford to be aligned with anybody like Senator Johnson.

The feedback from the group was forceful . . . *anybody but Johnson*. He was a Republican right-winger. He caused too much trouble as it was.

"Do you see anybody else returning your calls?"

Ken had a point. And Bri had started to trust Ken's advice. He was becoming a much-valued mentor to Bri, coaching her as she eased into the more public elements of this new role as vaccine advocate, and already pushing her beyond her comfort zone.

They arranged a group call with Senator Johnson. They had no choice but to get political.

CHAPTER 7

GETTING POLITICAL

There were over a hundred of the injured on that group call with Senator Johnson, most of whom were extremely skeptical about him, and very reluctant to be associated with such a controversial figure. Nobody really had any idea how the meeting would go, but given how they had all been dismissed by multiple medical professionals as well as their own senators, and given what the media was saying about Johnson, it was generally assumed that the meeting would not go especially well. They'd probably just spend the time getting bossed around and gaslit by this crazy right-winger, who somehow wanted to use them in order to feed his own ego and his own agenda.

But he was the only one willing to meet with them. And they were desperate. So they planned a meeting with him, choosing a "safe word" as a way of communicating with each other if they needed to just end the meeting and bail.

Senator Johnson brought a doctor on the call with him, who immediately took control of the meeting, claiming that it was impossible to get Covid twice, and showing little interest in talking about vaccines. Everybody started messaging Bri the safe word. Her phone buzzed and buzzed as the other injured expressed their concern at the direction the meeting was taking. Bri had to jump in.

"Excuse me for interrupting, but this is not the purpose of this meeting."

The doctor ignored Bri, and continued with what he wanted to say, until Senator Johnson cut him off, saying that Bri was right. Johnson turned the floor over to the injured, and the doctor abruptly left the meeting. Now, the injured could speak, but would they be listened to?

They spoke about their injuries, their treatment, and their everyday battles, as well as the impact that the entire experience was having on their families, their finances, and their feelings about the future. They spoke about how they had tried to get help, but none seemed to be there. Anywhere. They spoke about how their own senators—both Democrats and Republicans—weren't interested. They spoke about how *nobody* was listening to them.

But Senator Johnson was listening.

He not only listened, he took notes too; the sound of his pen scribbling away on the paper could be heard in the background as one person after another opened their hearts to him; for many of them he was the first person in authority to even listen to them. He treated each of them with kindness and respect. Johnson shook his head in places, looking up from his note-taking to offer sympathetic words that seemed to come straight from *his* heart. He stayed for over an hour beyond the meeting time inviting anyone to share, who wanted to share.

The injured got on a chat together after the call finished, with all of them surprised at how the meeting had turned out. They were especially surprised to find what a *gentleman* Senator Johnson seemed to be; he was not at all like the person they'd heard about in the media. He didn't seem like a crazy right-winger. He seemed like a rather gentle, kind, and concerned man who wanted to do the right thing. Perhaps the media was right about some things, but how could reporters miss this side of him? Senator Johnson's kindness won over the vaccine-injured.

At Ken's request, Johnson agreed to host a press conference in Milwaukee toward the end of June—they had just a few weeks to get themselves organized.

Ken took the lead to organize the conference. On the agenda were Sheryl, Candace, Kristi, and Bri who would each tell their stories, and Steph to tell Maddie's story, with now-thirteen-year-old Maddie accompanying her. Danice couldn't get to Milwaukee, despite what a vital role she was playing in all the advocacy work. The others pleaded with her to join them—Dance was a much-loved and respected member of their little group, but she was too frail to travel, and the fear of Covid was too great given her physical condition. Life was complicated and challenging enough especially for those who were speaking out, and Danice couldn't face her own life getting more complicated and challenging with a risk that she felt could possibly be life-ending. She'd continue advocating and working for the community, but from here on, it would be in the background.

Bri initially didn't plan on going either—she'd had the NIH study trip planned for exactly the same time, and also preferred to be in the background, so she told them all to go ahead and do the press conference without her. But Ken wanted Bri there—it was essential that the world got to hear about her experience as a clinical trial participant—so the date of the conference was moved to accommodate Bri's trip to the NIH in Maryland.

Ken organized everything for the press conference, planning for it to be held in the Packers' stadium, but the Packers and the NFL brand wouldn't allow it, so it would instead be held in the county building where Johnson had his office. The senator would introduce the event, Ken would introduce the speakers, and then it would be handed over to the injured, followed by a Q&A. Ken gave each of them seven minutes to tell their stories, with Bri going last. Bri contemplated . . . *seven minutes? How do I talk about my reaction, clinical trial problems, therapies that might work, as well as everything else . . . in seven minutes?*

Ken coached them all in their presentations, ruthless in his editing to help the women convey the key points of everything they had been through. The editing was vital. Their experiences were so multi-layered, and so *very* traumatic to go through in private settings, let alone a public one with cameras pointing at their faces, knowing that their every word would be on record for years to come. They all knew that once they did this and their stories were broadcast all over the world that international investigations would start, and their lives would never be the same again. It was a very daunting prospect, and "Coach Ken" kept the women focused on what they were saying, rather than overwhelmed by what they were *feeling*.

They decided not to mention the NIH to Senator Johnson or as part of their presentation—still very keen to protect Dr. Nath and his team by keeping quiet about the study, and still a little nervous about being aligned with Johnson—but Bri did let Dr. Nath know about the forthcoming press conference. He commended Bri and her friends on their advocacy and asked her to ensure she mentioned two things in her presentation: firstly, that their conditions *were* treatable; and secondly, that early intervention was key to recovery. Bri agreed, grateful for the advice, but did wonder why someone as senior and connected as Dr. Nath—who had a direct line to President Biden's chief medical advisor, Dr. Fauci—wasn't giving his own press conference to tell the world that himself. *Maybe he's waiting until the study is completed, and then he'll do a proper press release that the medical community will finally listen to!*

Suddenly, the speeches needed to be cut down to five minutes each rather than seven . . . then four minutes, then three minutes; a fifteen-minute press conference had more chance of being attended than the original thirty-five-minute one. And they could have breakout sessions where the reporters would be able to speak directly and individually to them all. There would still be plenty of opportunity for the press to get the full understanding of what was going on.

Nonetheless, Bri and her friends were disappointed. They were representing so many people, and all they were going to get was three minutes each? There was a lot involved in getting themselves to Milwaukee—they were all paying for the travel costs themselves, and they all knew they'd pay for it with their health afterwards too.

Bri had started to see an improvement in her own health since day three of starting IVIG. She had had her first sessions at the NIH, just the week before the Milwaukee press conference. She had got on the plane to Maryland being pushed in a wheelchair by Brian, and lay horizontal for the entire flight; whereas on the way home she walked herself on to the plane, and sat upright during the flight. *Maybe I AM going to get better!* Her body felt less heavy, her legs that little bit easier to move, her thinking was clearer, and she could sleep for an entire night—the first time in eight very long months.

Brian set up a daybed in the living room so that Bri could feel more able to participate in family life again. She propped herself up on the pillows, thrilled to not be stuck in a room alone all day and night with the physical and emotional darkness; grateful to be able to tolerate the sound of her children who were now thrilled to have their mom back in their environment again, even if she was confined to a daybed.

The past eight months had taken its toll on Cooper and Hannah, who had responded to their mother's illness—and absence from their lives—in different ways. Hannah had taken to writing her mom little love notes that she would push under the door to the bedroom that she was supposed to stay out of. Bri would often find pieces of pink paper folded up, with the words, "I love my mommy!" or "You are the BEST mommy in the whole world!" surrounded by little hand-drawn hearts.

Cooper, however, had become angry and afraid, and refused to go to school; instead, needing to know where his mom was at all times. His little shoulders had become tense; his body hunched up. If she was in the spare bedroom, Cooper would wait until he thought she might be asleep, then push the door open just a little bit so he could peek in and watch his mom sleeping. If she wasn't there, he would frantically search the house until he found her.

After the IVIG treatment started, Bri was still in a lot of pain, but the pain was just that bit less than before—creating space for her to at least *pretend* she was fine to her kids; to pretend that Hannah snuggling up to her on the daybed didn't hurt anymore; to pretend that all she wanted to do was some coloring too. She just wanted everything to be normal for her family again, and lying on the makeshift bed in the living room felt like it was a tiny step in that direction.

The IVIG treatment gave Bri the capacity to pretend that everything was going to be OK.

It also gave her the capacity to become more proactive in addressing the trial clinic, asking them where the promised reimbursement of her medical expenses were and reminding them that AstraZeneca *still* hadn't been in touch.

The lack of acknowledgment from the much-celebrated vaccine manufacturer in Oxford was making it harder and harder for Bri to confidently state something of which she had previously been so proud . . . "I'm pro-vaccine." AstraZeneca had now ignored her for months. That didn't feel like very "pro-vaccine" behavior.

Does being "pro-vaccine" mean ignoring people who are harmed by vaccines? Surely the most pro-vaccine thing you can do is care for those that are harmed by them?

Bri had to think carefully about what being "pro-vaccine" even meant to her. Did it mean believing in vaccines? And what did that mean . . . "believing" in them? Did it mean promoting vaccines? Getting vaccines? Encouraging others to get them? Did it mean believing that vaccines were the *only* way of dealing with certain health issues, and that other ways shouldn't even be considered? What exactly had she meant all these years when she had proudly considered herself pro-vaccine? What did *anybody* mean when they claimed to be pro-vaccine? She had never so much as read a patient information leaflet when she'd got any other shot, let alone read any actual clinical trial data. She didn't even know how clinical trials were run before this. She didn't know that vaccines were sometimes tested against *other* vaccines and not saline. She'd had no idea what happened to anyone who had an adverse reaction. She'd never had a conversation about vaccines with her doctor for anything more than a few minutes before rolling up her sleeve before.

Had she really, *ever* been pro-vaccine? Or had she simply been vaccine-ignorant?

Every one of them would call themselves "pro-vaccine" during their presentations for Milwaukee; it would be stated by Senator Ron Johnson in his opening remarks, and by Ken in his introduction of the group. And across the globe, at some point or another every single person injured by the Covid vaccine would have found it necessary to state, "I'm pro-vaccine" before they felt they had even the slightest chance of being taken seriously by a medical professional, a politician, and even their friends.

Bri struggled to think of any other medical conditions where people felt the need to state their personal health beliefs before they were listened to. Vaccination—something that an individual made the decision to accept into their own bodies in order to maintain their health—was actually a much more complex and highly controversial topic involving multiple layers of society and numerous *government* bodies.

The fact that her illness had been caused by a vaccine meant that her personal situation was actually a very political one.

The entire injured community felt the politicization of their health too. Their support groups were infiltrated by people who *hadn't* been injured, at least not by the Covid vaccine. Some of the infiltrators were people who had been injured by other vaccines in the past, and they were understandably angry at the years of being dismissed and discarded by their doctors, families, and friends, in the same way that the Covid vaccine-injured were now being. Some of them weren't injured themselves, but had spent years educating themselves about vaccines and had tried educating others, again, to have their voices discarded and dismissed. Some of them were people who were angry about the lockdowns and looming mandates, concerned about the control that had been imposed over the country since Covid. Some questioned whether Covid existed.

They descended upon the support groups for the vaccine-injured and brought their anger and frustration with them, which they shared with some incredibly scared and vulnerable people. The largest of the support groups changed its name to "A Wee Sprinkle of Hope," to reflect the culture of compassion the admins were committed to creating, as they set about removing anyone from the group who wasn't actually Covid vaccine-injured, or who was contributing with anything but kindness. Many of the injured felt like they were dying; barely clinging on to life. The people in these support groups needed love, not hate.

Those who took on leadership roles within the vaccine-injured movement initially had no idea just how political their efforts would ultimately become—all they wanted to do was help people who were suffering. Many of them had a history of helping others, just like Bri with her efforts to cheer up her community during the months before she became injured, or setting up her preschool before that, and even in the years before that when she voluntarily mentored teenage girls through a program at the church. All the admins of these support groups—themselves injured—wanted to ease others' pain, but for many in leadership roles, that would mean getting involved in the politics of something they were only just beginning to comprehend.

* * *

Sheryl and Ken, Steph and Maddie, Candace, Kristi, and Bri helped themselves to the buffet that Senator Johnson had laid out before the press conference. Bri was still struggling with food but nibbled on some fruit as she lay stretched out across several chairs, with her head in the lap of the friend who had accompanied her.

Bri had been keeping her distance from Johnson a little—he had been reported in the media *again* recently about his ideas about vaccines, and she still wasn't entirely convinced that they were doing the right thing in being aligned with him. The senator took one look at Bri lying horizontal and immediately asked if she was OK; whether there was anything she needed in order to be more comfortable. She explained that she needed to lie down as much as possible—even at the airport she had laid down on the floor next to her friend because she couldn't even sit upright in her wheelchair for long. It was always so embarrassing; lying on the floor all skin and bones.

The senator asked his team to arrange for a bench to be placed behind Bri's seat at the conference table. He spoke to his staff kindly and respectfully; the difference between how he behaved in-person compared with how the media presented him astounded not only Bri, but the whole gang. He spoke quickly, with a thick Wisconsin accent—that part they all knew from TV. But he listened . . . *really* listened . . . paying attention to what was being said to him and being kind and respectful in his responses. He always seemed to know the right thing to say to his physically and emotionally fragile guests, and his team were the same.

Bri and Maddie peered out of the windows as the reporters entered the building—ten or twelve of them with huge cameras, tripods, microphones, and other recording equipment.

"Holy shit," said Maddie, eyes wide at all the cameras. Her mom immediately told her off, "Shhh! That's not OK!" thinking that was all they needed, a teenager with a potty mouth in front of all this media. But Steph was thinking the same thing. All the adults were. The enormity of what they were about to do was not lost on any of them.

Johnson prepared them for what he thought was going to happen—he would give the reporters some time just with him first so they could pull him apart, which is what he was used to. He said that the reporters would likely be focused on getting some quote out of him that they could then splash all over the media out of context. *Wow—so he knows he's going to get abuse for this but he's still willing to do it.* Then the injured could tell their stories.

He hadn't asked to see their speeches beforehand, nor had any influence over what they wanted to say. On the contrary, the group had told Johnson that they would *only* talk about their injuries, and were not going to address any of the topics that Johnson had been quoted in the media as talking about before, such as alternatives to vaccination for dealing with Covid. He had agreed to all their terms and asked for nothing in return, giving them a completely open, free platform to tell their stories to the media, who would then share their stories with the world.

It was go time.

They entered the conference room one by one, with Maddie's eyes getting bigger and bigger at the sight of all the cameras, the microphones on the tables, and the little sign with her name on it in front of where she was going to sit. The cameras started clicking. And Bri could feel her own eyes getting bigger as well as her heart pounding and mind racing *What if I pass out? What if I lose my place on my speech? What if? What if?*

Breathe, Bri, breathe. This is just like cliff-diving at Lake Powell.

And then, at last, the silence was broken. They told their stories; the first time they had said them out loud, with the trauma and tension that had led them to that moment spilling out along with their tears. Steph's pain was especially evident, as she expressed the agony of a mother who listened to her daughter begging for relief from the pain, while the doctors had turned the other way.

Within fifteen minutes, they were done. Bri looked to Coach Ken for reassurance as she finished her speech—it had been years since she'd done any kind of public speaking. He nodded his approval at her. She'd done brilliantly. They'd *all* done brilliantly. They'd done the right thing by sharing their stories with the media; somebody else would take responsibility for the vaccine-injured now; at last, the Covid vaccine-injured in America would be taken seriously, and taken care of.

"Who paid for your travel?" "Are you going to sue?"

The questions took the group off-guard. They had expected to be asked about their symptoms, their doctors, their treatment, maybe even their thoughts about the Covid vaccine in general; they hadn't expected to be asked about money. None of this was about money. Sure, they were all spending a fortune on their own health, but *money* wasn't the reason they wanted to tell the world about what was happening to them. If any of them had been given a choice—to have money or to have their health back—every single one of them would have chosen their health.

They wanted to be seen, they wanted to be heard, and they wanted to be believed. They wanted these things for themselves, but mainly they wanted them for a community that was also suffering and depending on them to break through a wall of silence. They were doing this because it was their duty.

The proceedings came to a close, and the injured left the conference room. The reporters hadn't wanted to speak directly to them so there weren't any breakout sessions after all. Instead, they flocked to Senator Johnson, asking him how he felt about parading a bunch of sick people for his own agenda. Johnson had been expecting it. It seemed that the media had a certain image of Johnson that they were keen to reinforce.

Nonetheless, Bri and the others were elated. A huge weight was off their shoulders. All their efforts were going to pay off—two of the reporters had been spotted actually *crying* during the conference. They'll definitely make a huge deal about everything. The world will hear what's going on. They will demand reviews and investigations. There will be outrage. Commissions to help the injured will follow. Research and directives will be provided to the medical community. The injured advocates could all take a step back now. The hard part was over.

Now they could focus on the easy part—the beautiful and most unexpected joy from becoming vaccine-injured—the friendships that were developing between them all.

* * *

There was a very special bond between the injured; it somehow felt like they'd known each other their whole lives even though they'd never actually met in person. It wasn't just the fact that they were sharing the same experience—an experience that was culturally taboo as well as personally terrifying—but it was as if they shared the same souls. They'd greeted each other with huge smiles and warm hugs; keen to embrace their new friends as if to envelope them in whatever healing their own hearts had to spare.

Maddie and Bri in particular had hit it off immediately . . . not only did they have the fact that they were both clinical trial participants in common, but Bri had a rather childish sense of humor that teenage Maddie shared. They were always giggling together about things they managed to find amusing about their situations, like inconvenient and embarrassing times they'd peed their pants, and sometimes made each other laugh so much that another pants-peeing incident was always imminent. After months of being confined to a bed, of missing out on school, of not hanging out with her friends, of losing so much of her *life*, Maddie at last found somewhere that she belonged. There were others like her; she wasn't alone.

Maddie's mom had also been greatly encouraged by the presence of these new women in her life; women near her own age yet suffering in the same way as her daughter. The NIH had stepped in and told Maddie's doctors that they had to start taking Maddie's condition seriously, which they did, and Steph had started to feel more optimistic about the future, despite being deeply traumatized about the past.

They all spent the entire time in Milwaukee together, laughing and joking and doing tricks in their wheelchairs. A bystander would have had *no idea* what the group of new friends were going through, as they high-fived each other for *not* tipping over on the sidewalk. Even Candace, who was persevering with her walker rather than a wheelchair, would simply shrug when her ankles rolled over, almost bringing her to the ground.

They kept the conversation away from the heaviness that their lives had taken on since their vaccines, and instead delighted in making one another smile.

They explored the city a little, with Bri admiring the beautiful old architecture and stained glass windows, frequently stopping just so she could get herself horizontal and rest on one of the park benches they'd come across. Finally, she decided that she just had to get back to her room, and laughed when everybody insisted on joining her, with Maddie climbing on to Bri's bed with her as if she'd known her forever. The little girl chatted away like she was at one of the sleepovers she used to love so much, talking about how *totally embarrassing* it was when her mom cried in front of all those cameras that morning.

They turned on a local TV channel, curious to see if any of those cameras had got their story out yet. Most of the reporters had been from local news outlets, with a couple attending because Ken was a celebrity to local sports fans. Even if the story hadn't yet reached the national news, there might be something on the local news first. Nothing. They flipped through the TV, stopping on any news channels. Still nothing. Maybe it was too early in the day. They'll be something on later. *They must be saving it as a big story for this evening's news. It's going to be something the world has never heard before!*

But all they could find was a very small mention of the press conference on Fox News that evening. *Is that it? OK, tomorrow . . .*

But the next day it wasn't headline news. The only news that could be found about the conference were reports of Senator Johnson organizing a misinformation panel in his efforts to *yet again* mislead the public about the Covid vaccines. That was it.

They were all stunned. The electric shocks that raged through Bri's body were not misinformation. Sheryl's loss of cognitive functioning was not misinformation. Candace's walker was not misinformation. Kristi's internal vibrations were not misinformation. And Maddie's feeding tube was not misinformation. The fact that every single one of the vaccine manufacturers had abandoned them was *not* misinformation. What was happening?

Where was the concern? Where was the outrage? *Where was the humanity?*

The reporters didn't share the stories they had witnessed that day. Senator Johnson had been right; the journalists *had* already decided what the story was going to be, and that story would be centered around a crazy right-winger and the latest circus he was ringleader of. Within hours the local medical boards issued their own statements criticizing the senator, and the local doctors went after him.

In that moment, Bri realized exactly *why* she had believed what she believed about Senator Johnson, prior to meeting him. She'd never met him before. She'd never had any interaction with him. Her beliefs about him had been based *entirely* on what the media had wanted her to believe. And for some reason, the media wanted people to believe that Johnson— the man she had found to be kind and compassionate—was instead deranged and very dangerous.

Instead of reporting on the vaccine-injured, the media had reported on a "crazy" politician. As if the vaccine-injured didn't even exist. And if the vaccine-injured didn't exist in the media, then the vaccine-injured didn't exist at all. *Is that what the media wants people to think? That we don't exist?*

The questions and realizations whirled around Bri's tired brain on the flight all the way back to Utah. She didn't know who or what she believed in anymore. The events of the past eight months had proved to her that the institutions that she had always believed were honest and trustworthy with a sense of humanity and justice at their core, were perhaps not any of those things. She had been let down by the healthcare system, she had been let down by the political system, and now she had been let down by the public messaging system. After all the effort that she and her friends had put in to patiently, respectfully, and collectively advocate, their voices were not going to be heard. All this time, they had assumed that they just needed some media coverage, and the battle would be over. It was a shock to find out that this was not the case at all.

At least they had two things they *could* still rely on: the largest social media platform in the world; and that all-important and much-valued part of the Constitution guaranteeing all Americans the right to free speech. They still had Facebook and the First Amendment.

Didn't they?

CHAPTER 8

GETTING CENSORED

They'd all assumed that the press conference would be the end of the road for their advocacy, but instead it turned out to be the beginning. Bri's own senator, Mike Lee, had teamed up with Senator Johnson and together they sent a long letter to the FDA, requesting answers to some very specific questions about the clinical trials, and wanting to know what the CDC had in place for dealing with those adversely impacted by the vaccines. They deliberately made the letter public.

Fox6 News Milwaukee had streamed the press conference live on YouTube, and the entire event remained on its channel, eventually gathering well over a million views. The speakers were Googled by people all over the world, many of whom were themselves injured. Until then they had assumed that they were alone.

Bri contacted her state's health department and was surprised to learn that the vaccine director there had himself had a scary—albeit temporary—reaction to his own vaccine. He told her that if these reactions were happening to others, then he needed to hear it from the federal level. She asked Dr. Nath to get involved, but he wanted to wait until the NIH was ready to go public themselves about the reactions. The state health department stopped talking to her. *They think I'm making it up. Why won't Dr. Nath say anything?*

There were plenty of others talking to Bri and her injured friends, some of whom were very angry and felt the need to tell the injured that they were all going to die, and they deserved it. The private messaging and public commenting about how the stupid vaccine-injured "sheep" had chosen to be poisoned were intensely distressing to people who spent every day feeling *exactly* that. Stupid, and poisoned.

And then there was another kind of anger, this time from people accusing Bri and the rest of them for being liars and murderers. Liars, because if anything they said was *actually* true then it would be all over the media, with the government taking action. Murderers because they were directly responsible for people declining the Covid vaccine, and those people would end up dying. Bri and her friends were killing people, and they had to be stopped.

If only the NIH and FDA would tell them what they knew, then maybe they wouldn't flood us with this hate. Why won't Dr. Nath tell them?

Bri focused on the countless new injured that were reaching out and directed them to the online support groups—reassuring them that they absolutely were *not* alone. They could find the support and friendship that had been sadly lacking in so many of their lives since they'd become sick. The new name for the largest group—A Wee Sprinkle of Hope—was very appropriate. Hope was exactly what was needed for some very desperate people, and the online support groups, quite literally, kept people alive.

Then, twenty-four hours after the Milwaukee press conference, Facebook started shutting down those groups.

The first to go was one of Bri's support groups. Its members ran around the social media platform, desperately searching for any other injured with whom they hadn't connected privately. Somebody knew this person, who knew this person, who knew this person. Anybody that was known to be suicidal had to be found. Nobody could be left behind. *Get everyone over to Wee Sprinkle.*

Within days, the injured gathered together, making the largest group even larger. The Wee Sprinkle admins were completely overwhelmed—frantically checking and double-checking who was actually injured and in need of help, and who was trying to get in because of ulterior motives. Admins of the smaller groups told all their members to join the biggest one, just in case they got shut down too. Within a week, most of the injured were in A Wee Sprinkle of Hope. There were thousands of them. With more joining every day.

Then, five days after the conference, Facebook—without warning—shut it down. It was the largest Covid vaccine injury support group in the world. And it was shut down on the exact same day Bri made her first post in the group.

The injured froze. What was happening? They couldn't talk to their doctors about their health. They couldn't talk to their senators about it; many of them couldn't talk to their friends or their families. *Now we can't talk to each other?!* In closing down the support groups, the social media platform was actively restricting people in horrific physical and emotional pain from communicating with each other—isolating them so that they had to deal with their suffering alone. And by doing so, sending the very same message to all of them that Bri had consistently received from the clinical trial clinic and AstraZeneca: *you mean nothing.*

Is there any other medical condition whose sufferers would be censored like this?

Many of the injured refused to give up, and instead learned how to play by Facebook's rules. They self-censored, developing code words so that they could continue writing about their struggles and symptoms online. "Carrot" or "the gift" was used instead of "vaccine." "Dancing" became a way of saying that you'd been vaccinated . . . *I've danced once, I've danced twice, I've never danced but my sister got a blood clot in her leg after the second time she danced* . . . until a mainstream news article exposed a "dance" group with nearly 30,000 members, and that one got shut down too.

When Bri went back to the clinical trial participants' support group she had joined on the day of her shot, she found that every single post where people had talked about their side effects was *gone.* It had always been obvious that the moderators promoted vaccination but now, anything that could possibly be construed as talking about the vaccines in anything but a positive light had been completely wiped off the forum. *Was that the moderators? Or Facebook? What happened?* Bri assumed it was because they didn't want their forums to put anyone off participating in clinical trials—without clinical trial volunteers, then there wouldn't be any vaccines at all.

Warnings appeared underneath people's posts on their own pages about their own injuries; the warnings urged the viewer to go to Facebook's Community Guidelines to get accurate information about the Covid vaccine. The warnings also deterred other Facebook members to *not* interact with the person who they considered to be a repeat offender of posting misinformation. The person posting had no idea these warnings had been attached to their post until a friend would alert them to the fact.

Some of the injured discovered that their posts weren't even being seen by their online network of friends—injured or non-injured. It became apparent that some people were being "shadowbanned"—their posts were being hidden by Facebook algorithms so that they wouldn't appear on friends' feeds or would only appear if the poster's actual name was typed in the search bar.

Some of them found other platforms such as Telegram, through which they could communicate more freely, but Facebook was still the most effective way of connecting with people, and had become the only connection with the outside world for many people whose normal lives had been confined to four walls and a bed after becoming injured. Social media could break through those walls and because of that, possessed immense power. It was a lifeline.

The censorship killed that lifeline, and left people feeling even more desperate, as Bri would find out. Soon after the groups were shut down, there was another suicide.

Amid the chaos of people losing their only source of support, one woman had reached out to multiple public figures in the vaccine-injured community, including Bri. She had been posting cries for help on her private page for months, growing angrier and angrier as the algorithms continued to dramatically reduce the visibility of her posts. Eventually she resorted to sending long private messages, clearly articulating just how desperate she was. All of the messages remained unread. When the news of her suicide reached the other injured, they scrolled through their private messages for any hint of just how close to the edge she was. Nothing. There were no messages. *How could she not have told us?* And then one of the admins actually typed her name into the messages' search bar, and suddenly all the cries for help appeared. *She had sent them to everyone.* Her cries for help had gone ignored, not because nobody cared, but because her suffering had been deemed unimportant by Facebook. She left behind a husband and four children.

* * *

Despite the online censorship, there was hope that some of the mainstream media would cover the stories of the vaccine-injured. Bri's story was aired on ABC4Utah, and a few days later on KMYU, another Utah-based TV station. On the same day as the KMYU story, Bri received a phone call from the clinical trial's manager of operations. It was the first time she had heard from the company in months.

The Dressens were told that a payment of $590.20 from AstraZeneca would be deposited in their bank account. Brian suggested they might be missing some zeroes—the amount was nowhere near what Bri's injury had cost them so far. A few hours later, the $590.20 appeared despite Brian having made it clear that they would only accept the payment if AstraZeneca confirmed that more was to come. AstraZeneca didn't confirm anything, and Bri wasn't to hear from the clinical trial company again for more months. *Maybe they're dealing with all the vaccine-injured in the UK?*

Bri had heard about a few injured based in the UK in the spring of 2021, but that summer they came in *droves*. One day after another she heard from someone else who was suffering after having had the AstraZeneca injection—the company had provided the majority of Covid vaccines in the UK, and word was spreading about how one of their clinical trial participants in the US had been injured and then ignored.

Bri found the sheer numbers of injured coming at her from the UK difficult to face at first. Many of them had been vaccinated during March or April, months before Bri herself had "come out." She wondered, if she had been louder earlier—if she had been braver—could she have prevented all this agony across the Atlantic? *Was that person who accused me of being selfish for not speaking out back in November actually right?*

She couldn't spend too much time with those kinds of thoughts. Her mental wellness had become a priority to her—it was key to being able to cope with her physical illness. And in a weird way, the Brits started to make Bri feel less alone. With AstraZeneca not being authorized or approved in the United States, she didn't know anyone who had had that particular brand of vaccine except another trial participant, Sarah Lee, and Sarah had remained very, very quiet about her reaction once she knew just how political everything was getting. The presence of the UK-based injured in Bri's life somehow brought her comfort; she felt like she would always have a home there.

* * *

The online broadcast of the press conference with Senator Johnson was still reaching people all over the world, and they continued to contact Bri and the other speakers. New support groups were being set up, all with very discreet names so that it wasn't entirely possible to know upon first glance what they were for. It made it incredibly difficult to be seen by those who needed support but minimized negative attention. Members had to answer strict questions before being allowed in, and it took a while for them to learn all the code words. Nobody would have known these online groups existed unless they were directly introduced to them. People all over the world still thought they were suffering alone.

Optimistic and persistent, the injured advocates reasoned that just because the conference reporters hadn't covered it, that didn't mean that others wouldn't. And they still had the reporters from Pittsburgh, the *LA Times*, and *New York Times* on their side.

But then the Pittsburgh reporter told Kristi that she had asked for statements from the drug companies, and a senior person at Pfizer had returned her call, questioning her integrity as a journalist for even considering writing about vaccine injuries, and threatening her career if she went ahead with it. She then stopped speaking to the entire group. At the same time, a Utah reporter had been threatened by the state health department, and the two major hospital systems in the state, effectively telling her to "shut up, or else" if she continued to report on vaccine injuries. She was so sorry, but she couldn't carry on.

The *New York Times* had originally planned on their article to be centered around Heidi's suicide, and had interviewed her grieving husband at length, as well as Steph. Bri sent the journalist scientific articles that Brian had found about adverse reactions to vaccines, and introduced her to Dr. Nath, explaining that the NIH was working with them and would be able to corroborate everything. But then the journalist said that she couldn't publish the story after all; they weren't allowed to write anything that would make the vaccines look bad.

The *LA Times* did run a story, but the quote attributed to the NIH seemed to minimize the issue, stating that cases of injury were extremely rare. When confronted, Dr. Nath said he'd been misquoted, but the journalist fiercely defended his professionalism. Danice told the others that something wasn't quite right. Maybe this is what Casey—the NIH employee—had been trying to tell them. *Journalists misquote people all the time. Just look at all the effort the NIH has put into getting us access to treatments! All their emails are saying they're going to help! It's the media messing everything up. Or is it?*

The *Wall Street Journal* was ready to go with the story but just needed a quote from Dr. Nath. The journalist had been regularly calling Bri to clarify one thing after another, and then suddenly he stopped calling. Three weeks later Bri finally heard from him again, but this time it was quite a different conversation. He had had a long conversation with Dr. Nath, who had explained just how dangerous Covid was, and how dangerous it would be to run with a story that could put people off having the vaccine.

All this time they had been working so hard on connecting with the media, spending countless hours reliving their own horrific stories, and taking time away from their own healing in the hope that it would help others with theirs. All this time they had been sending journalist after journalist to Dr. Nath to corroborate their stories and confirm that the NIH was working with vaccine injury survivors. And the journalists had just kept disappearing on them. *Has the NIH been putting a stop to our stories, all this time?*

The NIH team still hadn't completed the study, but they had managed to publish a paper in a medical journal specifically in response to videos that had been circulating online of people with shaking and tremors post-vaccination. The paper had emphasized the safety and efficacy of the Covid vaccines, detailed case studies of people who had experienced disturbing neurological symptoms post-vaccine, and stated that the patients had FND (Functional Neurological Disorder); FND is a disorder for which no apparent cause can be found, so is assumed to be related to anxiety, stress, and past trauma. The paper assured the reader that the Covid vaccine was not responsible for such neurological symptoms. Danice had confronted Drs. Nath and Safavi with how dangerous it was to effectively be telling people that these terrifying symptoms were all in their heads or all of their own making, and the doctors had reassured Danice that that was not their belief.

Nothing made sense. The NIH *was* working with the injured community and had been for months now, flying countless of injured out to Maryland, validating all of their symptoms, not just testing but actually treating them, and then guiding their healthcare teams on how to continue that treatment. *They'd even taken Heidi's body parts for research.*

Toward the end of that summer, Bri and Danice were getting mixed messages from the injured they'd introduced to Dr. Nath and his team. Firstly, some people were rejected from participating in the study, whereas before the NIH had welcomed everyone. Then, some of the injured were being told that there wasn't even any research going on at the NIH. Dr. Nath stopped responding to anybody's emails, and the FDA stopped communicating with Danice. Something wasn't right. Had Casey's suspicions been accurate after all?

Bri managed to get Dr. Nath on a follow-up telehealth video call, but this time he seemed different than he'd been on previous calls. It was as if he couldn't look her in the eye. That same month, the NIH shut down the entire study, canceled Bri's second trip to the NIH, and left the vaccine-injured without any federal response to their conditions.

Bri had spent eight months telling the other injured to hold on; that the NIH were there for them. She had personally introduced over a hundred of the injured to Dr. Nath. She had done everything the "right" way, not wanting to make a fuss. And people had trusted her. She would never forgive herself for unintentionally leading them on.

Bri may have been doing her best all this time to *not* make a fuss, but others had thrown caution to the wind, to their own detriment. Some of the Disinformation Dozen—including Robert F. Kennedy Jr.—had their social media accounts shut down or restricted, and the surgeon general issued a warning against misinformation as the government started working more closely with social media platforms to control what was and wasn't said about the Covid vaccines.

It wasn't just the doctors, authors, and activists who were censored or discredited. Celebrities and other public figures were too. Musician Eric Clapton spoke publicly about the adverse reaction he'd experienced after both his AstraZeneca shots, only to find his drunk and racist comments during a concert back in the seventies to reemerge in British news; behavior he had long ago apologized for. Novak Djokovic, arguably the best tennis player in the world at the time, was ridiculed for his refusal to reveal his vaccination status, prompting the mainstream media to run articles calling him "whacky" and "eccentric." Green Bay Packers' quarterback Aaron Rodgers had his participation in the sport restricted after saying that he wouldn't be getting a Covid vaccine due to being allergic to one of the ingredients. And basketball player Kyrie Irving was banned from practicing or playing with the Brooklyn Nets unless he had the vaccine.

Any public figure refusing to have the vaccine, refusing to state whether they'd had one or not, or talking about the vaccine making them feel unwell, was abused online and ridiculed by the mainstream media. And that summer, the American Board of Emergency Medicine issued a statement, warning their physicians that certification could be revoked for not following their Code of Professionalism. The statement featured a link for people to submit a violation report. The timing of the statement made it very clear; any doctors questioning the vaccine were expected to keep their questions to themselves.

It seemed that even the government's lead health officials had to be careful with what they said. Robert Ray Redfield Jr., Director of the CDC for three years until his abrupt departure in early 2021, had expressed concern over the federal management of the pandemic, conflation of numbers, and restriction of alternative interventions. He had claimed that other measures could be more effective than vaccination and wanted to consider natural immunity, but he had been ridiculed. Two senior vaccine reviewers at the FDA—Marion Gruber and Phil Krause—both with decades of experience, questioned the acceleration of the approval process. Dr. Gruber was specifically concerned about the increasing evidence of the association between the Covid vaccine and the development of myocarditis (a type of heart inflammation). Gruber and Krause were removed from the approval process, then both suddenly resigned around the same time. What was happening to some of the country's most seasoned experts? Could *anybody* question *anything* about the vaccines? Was anybody safe from repercussions if they did?

It should have been no surprise to the Dressens when Brian lost a major promotion for a job he had already been doing for the previous year. His employer implied that Bri's illness was a reason why.

Brian had studied all of the clinical trial reports ever since they'd come out a few months earlier. Moderna's had been first, and he found Olivia's reaction in there, completely downplayed. Olivia cried when she saw it; despite realizing from the start that she meant nothing to them, it still hurt her to see it in black and white. The Pfizer report significantly minimized and/or outright lied about Maddie's reaction, and there was no mention of Bri's reaction at all in the AstraZeneca report.

Brian testified at an FDA hearing that October, stating that 266 participants from Bri's trial alone had discontinued because of adverse events, with fifty-six of them being neurological in nature without any further explanation given on what these patients experienced. His pleading for the FDA not to authorize the vaccine for children was picked up by the media, and he was unexpectedly quoted all over the internet. A Google search resulted in pages and pages of links about his testimony, in multiple languages. His employer would *not* be happy about Brian drawing attention to himself, and he knew it. Four hours later, after dealing with his employer's concern, the same Google search resulted in completely different results, with hardly any mention of Brian in any mainstream media. *Wait a minute . . . Google is censored, too?*

The Dressens had been interested in the clinical trial reports not just to see what and how safety signals were being addressed, but also to see what was being said about the efficacy of the vaccines. They had been watching what was happening in Israel, a country that had vaccinated most of its population as soon as the vaccines had become available, yet were now getting breakthrough infections. They wondered how long it would be before breakthrough infections became evident in the United States. *What do we really know about the efficacy of these new vaccines in such a short space of time?*

These were such uncomfortable questions for the Dressens to be asking themselves; questions about the media, government, science, pharmaceuticals, censorship, and wondering whether all of those things could *possibly* be related? They knew, from painful personal experience, that the institutions they'd trusted were either hiding the truth or outright lying about adverse reactions to the Covid vaccines; but surely they couldn't be lying about anything else?

When a few media reports hinted at the vaccine perhaps not being quite as effective as originally believed, the Dressens weren't at all surprised. CNBC published an article based on a CDC study showing that 74 percent of the people in a recent Covid outbreak in Massachusetts had been vaccinated with two doses. The article went on to encourage everybody to wear masks because even the vaccinated people could be spreading the virus, and the Dressens waited for the government to come out and admit that actually, there might be a bit of a problem with the vaccines, after all. But they didn't.

They just kept on pushing the jabs, without any effort whatsoever to address the damage the jabs were doing.

Anybody with questions was deemed to be the worst kind of human being . . . far worse than being racist, sexist, or homophobic. During the Covid vaccine rollout the worst insult that you could hurl at someone would be to call them an anti-vaxxer. And the newly vaccine-injured were terrified of being called one. It wasn't unusual for doctors' appointments to begin with a vaccine-injured patient tearfully saying, "I'm not an anti-vaxxer"; the fear of being considered one had to be addressed before the fear of what was happening inside their bodies could be. As if them having any questions about vaccines would mean that their access to treatment would be denied.

What *was* an "anti-vaxxer?" What did it even mean? Bri started wondering about this term that she herself had used in the past, but similarly to how she had used "pro-vaccine," she had never *really* thought much about it. It was an insult, but why? Why would somebody's opinions about vaccines be something to which the government, the media, and society in general felt the need to assign *such* negative connotations? And why were the vaccine-injured being called anti-vax? They'd had at least one vaccine in their lives. *How is it that people who have actually been vaccinated are being accused of being anti-vaccination?* Like so much of what Bri was learning, calling vaccine-injured people "anti-vaxxers" didn't make sense, whichever way you looked at it.

Was the insult just another way of silencing people?

Who else might have been silenced when talking about possible preventatives or treatments for Covid? Even before the vaccine had become available, there had been physicians talking about repurposing existing medications, about the benefits of vitamins C and D, about the use of copper and nicotine, and about natural immunity. There were social and health experts talking about how to manage illness independently of the healthcare system, but all anybody seemed interested in doing was telling everyone to stay at home until there was a vaccine.

Whether or not any of those strategies were effective, surely a global crisis warranted full exploration of *all* the possible solutions? Instead, they had all been dismissed, with the FDA warning about the dangers of "horse paste" and the media ridiculing anyone who talked about "fishtank chemicals." Under the guise of public health and safety, the vaccine-injured, anybody who supported them, and anybody that had even so much as suggested an alternative way of managing Covid, were all lumped together as being "anti-vaxxers." And anti-vaxxers didn't deserve a platform.

Free speech in America was over.

* * *

When Bri was a little girl, like all Americans, she'd learned about how the country declared its independence from British rule, and about how vital that independence and freedom was to every single American. To be an American meant to be free. It was the reason why, for hundreds of years, people from all over the world had chosen America to be their home. She was taught that the freedom America provided was to be honored and protected at all costs, and this was a big reason why her husband chose the work he did with the Department of Defense.

A cornerstone to American rights was the First Amendment: freedom of speech, freedom of the press, and religious freedom. Freedom of speech was especially revered. The Founding Fathers had known how vital "getting the word out" would be to inspire their countrymen and women to rise up in defense of their independence and freedom from British oppression. Hamilton and co. wrote over one hundred essays that were independently published and distributed, each one providing a little more information than the one before it, before finally winning over the people and moving them to action. Without it, the war would not have been won.

When Europe introduced "hate speech laws" around 2015, many Americans had considered the laws to go against everything that made a society free. Aside from anything that would incite violence, only dictatorships would make rules about what citizens could and couldn't say, however unkind, inappropriate, problematic, or downright offensive such speech might be.

American schools taught that the removal, or even restriction, of free speech and a free press would have dramatic repercussions for the entire nation. Back issues of *Stars and Stripes*, old comics, and pop culture journals were studied in classrooms across the country, along with Anne Frank's diary. The writing of the little girl forced to go into hiding with her family because of their religious beliefs had stayed with Bri, as it had stayed with millions of children for whom the book was a mandatory part of curricula all over the world. Teachers would ask their students what they would have done during those times; would they have spoken out against the unfair treatment of their fellow humans, and disobeyed authority at their own risk? Would they have secretly been supportive and done everything they could to help on a practical level, knowing the risks that would have been involved if their support had been uncovered? Would they have looked the other way? Or would they have sided with authority, and *attacked* anyone who wasn't toeing the line?

Every child's hand would go up at the first option. Everybody likes to think that they're brave enough to speak out for those whose voices have been silenced.

But then the class would learn about new concepts like "propaganda" and "censorship" and the role those things played in creating a society where some members of it are not free; not free to worship how they want, not free to love whom they want, not even free to move within society how they want. And the class would get a better understanding of how that had happened almost a hundred years ago. Still, to Bri and her classmates, the idea of all that happening here, in America, was impossible.

Until 2021, when Bri came to the conclusion that the press wasn't free at all, she wasn't allowed to have any opinion that differed from the government narrative, and pretty soon, she wouldn't be able to choose what to put in her own body.

The mandates were coming.

CHAPTER 9

GOING TO WASHINGTON

The idea of mandating people to get the Covid vaccine was intensely distressing to many of the Covid-vaccine injured. Some of them had become injured specifically because they had been instructed or pressured to get the vaccine by their employers, many of them in the healthcare industry. Now they were too sick to work.

The vaccine-injured watched the conversations about mandates spread throughout the country and on their social media. People would feel like they had no choice but to get the vaccine, under the assumption that the doctors and government would be there for them if something went wrong. Those already injured knew that wouldn't be the case and felt that other people needed to know this before they rolled up their sleeves.

Reports of hospitals being overwhelmed led Bri to wonder just how many of those patients were there because of a vaccine, or what role the vaccine may be playing in the staffing issues that some healthcare facilities were having. *We know there are so many more of us already out there.*

Public and private entities declared that they would expect their employees to be fully vaccinated. Some restaurants, stadiums, and other places of entertainment required proof of vaccination status; some educational facilities wouldn't permit unvaccinated students to enroll. Facebook was awash with people changing their profile picture to add a frame that proudly indicated that they were fully vaccinated, but the social media platform removed any frames that referred to being vaccine-injured.

There was a sense of the world opening up for those who were willing to comply. Being fully vaccinated would mean that you could go to work, out for dinner, or watch a sports game; and the more people who got fully vaccinated, the more the world *could* open up. For everyone. Except the vaccine-injured. Who didn't exist.

From their sickbeds, the vaccine-injured felt increasingly pushed out of society; left behind while everybody else moved on with their lives. They didn't just wonder about the kind of future their health limitations would now allow them, they also wondered about the kind of future *society* would allow them. There was a very real fear of America turning into a country where somebody could turn up at your door, needle in hand, and force it into your arm. And there would be nothing you could do about it.

It was tempting to just have that next shot anyway; more than a few of the desperate injured considered it—it could just be what they needed to finish them off and thereby end all the pain and misery.

At the same time as the idea and implementation of mandates was circulating in both the mainstream and non-mainstream media, it felt like the vaccines themselves were being even more actively promoted, with multiple stories running about how safe and effective they were. The take-up of the Covid vaccine in the USA hadn't been as enthusiastic as the government had hoped, and the mandates and media promotion were intended to combat what was now being called, "the pandemic of the unvaccinated."

The NFL ran vaccine webinars with football superstars, targeting fans and other players; Sheryl heard about them through Ken so she signed up, only to find FDA Director Peter Marks alongside some very well-known football players, answering questions that the participants posted in the chat. Sheryl typed in, "Dr. Marks, what do you think about vaccine injuries?" and "What about the NIH studying vaccine harms?" But her questions disappeared from the list of community questions on the screen. *Erased.* Instead, Dr. Marks reassured the large group of young and able-bodied athletes that the absolute worst that could happen was maybe getting some mild myocarditis and needing to go to the ER. But you'd recover with some Tylenol and rest.

The *New York Post* reported that the CDC had paid NFL alumni millions of dollars to promote the vaccine.

What little publicity the vaccine-injured were fighting so hard to get—often at the detriment of their physical and emotional health—was being drowned out by government- and celebrity-endorsed public messaging across every single media platform. The more the injured tried to speak out, the bigger the mountain it felt they had to climb, and the more they were shoved back down it, only to try to get back up again. Sometimes it felt like the mandates were in direct response to the injured's efforts. It was exhausting.

And now children were going to be vaccinated—which meant that more children like Maddie would be out there, somewhere. Despite Brian's testimony at the FDA, and the testimonies of other scientists and medical professionals now being quoted in some media, vaccinating children was going ahead. Nothing had been put in place to care for the injured adults, which meant that they also had nothing in place for the injured children who would come along. It was heartbreaking to imagine children going through the same pain and suffering that Bri was still experiencing, every single day.

Bri had managed to secure a meeting with Peter Marks himself, for which injured physicians including Danice prepared extensively. They were disappointed when only Dr. Marks' staff attended. His staff encouraged the attendees to send over their medical files, which the team would look into, but Danice clarified that the purpose of the meeting was to address the needs of the *thousands* who were now injured—and would be injured in the future—not just the few on the call. Mary told the FDA staff about the several vaccine death reports she had filed that had received absolutely no follow-up. She plainly pointed out problems with their reporting system: a medical professional had filed eleven deaths on VAERS, and only two of them had been followed up on, going against the FDA guidelines that stated that all deaths should be looked into. *This is not about our individual problems, this is about YOUR problem.*

Dr. Marks emailed Bri afterwards, saying that what the group was asking for was beyond the remit of the FDA.

Frustrated at herself for *still* believing that a government agency would want to help them, Bri had had enough. *The FDA doesn't even want to know we exist, let alone want to help us.* She sent Dr. Marks' email to Michael Baum, a lawyer with prior experience in representing cases of pharmaceutical harm, and Michael put Bri in touch with Peter Doshi.

Dr. Doshi was associate professor of pharmaceutical health services research at the University of Maryland School of Pharmacy, had a specific interest in the drug approval process, and was an expert in clinical trials. He was also a senior editor at the *British Medical Journal* (*BMJ*), in which he had been writing articles questioning multiple aspects of the Covid vaccine trials since the trials had begun. Like the injured, he was also being snubbed by the FDA.

Peter Marks may not have been interested in what Bri had to say, but Peter Doshi definitely was. Bri, Michael, and Peter Doshi got on a call.

Dr. Doshi had contacts with scientists who were also very interested in what Bri had to say, and he suggested that they organize a roundtable discussion. There would be experts analyzing the available science, and the injured could speak about their personal experiences. Between them, they would draw upon all their contacts and invite anyone that could facilitate change—other senators, health officials, and all forms of media—and also livestream the entire event. Whereas the Milwaukee press conference had been painted as an "anti-vax" event by a crazy politician, Doshi could pack this roundtable event with people who were highly respected. He wanted to bring in the best in the field and show everyone that the vaccine-injured weren't anecdotals.

Dr. Doshi gave Bri the encouragement that she needed to keep pushing on with the advocacy when it felt like all the doors were being slammed in her face. If Facebook hadn't shut down their support groups, if the NIH hadn't halted the study, if the media hadn't turned its back on them, if Dr. Marks had perhaps even *just* turned up to that meeting . . . maybe the injured wouldn't have considered going to Washington. It would take so much out of them, after all. But they felt that they had no choice. Nobody was listening.

Bri would coordinate the injured, Dr. Doshi would coordinate the doctors and scientists, and Michael would be one of the key planners. Peter brought in pharmaceutical policy expert Dr. Linda Wastila and pharmaceutical safety advocate and former FDA drug advisor Kim Witzcak, who helped with coordinating. They brought in other highly respected medical professionals and another experienced vaccine-injury lawyer, Aaron Siri. They were united in that they disagreed with the mandates, believed in informed consent, felt there was reason for discussing the limitations in the evidence base, and were concerned about the stifling environment for anybody who wanted to critically appraise the evidence. It would be a forum for the experts to fight for medical freedom and civil rights; the injured would be fighting for their lives. There would be personal and professional repercussions for all of them—including the doctors and scientists—and they all knew it, but it had to be done. It would be the first of many roundtable events that Senator Johnson would go on to host in Washington.

The roundtable planned on November 2, 2021, would give a voice—and validation—to the vaccine-injured that they had not yet been granted. Bri got organized.

Together with Danice, Kristi, Steph, and fellow support group leader Drew Franklin, Bri had been working on formalizing the vaccine-injured as some kind of official entity—they already knew that they needed something more than just their website. Drew, father of a new baby as well as a toddler, was a businessman from Ohio, with a background in biology, who had developed an autoimmune condition after his shot. He ran a post-vaccine autoimmune support group on Facebook and was keen to further research and foster more open and objective discussion in the scientific communities. He knew that a website featuring stories alone wasn't going to get the injured where they needed, to be taken seriously. He threw his hat in the ring.

Up until they'd met Drew, it had been just the women doing most of the advocacy effort on behalf of the injured; there seemed to be far more women than men who had been affected by the vaccines . . . about three times as many women. Women's healthcare was known worldwide to be dismissed more often than men's, possibly adding a further element to the difficulties that many of the injured were having in getting taken seriously. They were being prescribed anti-anxiety drugs rather than being fully investigated.

There were so many different injuries within the community affecting people quite literally from their head to their toes. Most people had multiple symptoms and—if they hadn't been told it was all in their heads—multiple diagnoses. It was important to get as many conditions represented as possible at the roundtable. They invited people who they thought might be willing and able to speak about their personal experience, knowing that if they agreed to come, the traveling and reliving of trauma would take a huge toll physically and mentally. There would be no benefit to their own personal health by going to Washington, they would be subject to abuse from multiple directions, and their reputations would be impacted forever, not necessarily in a good way.

But you don't do these things for yourself; you do them for others.

Steph and Maddie were up for it. Maddie's condition had been categorized by Pfizer as "functional abdominal pain" in their clinical trial reports. Steph had been deeply distressed to discover that the complex and debilitating condition that confined her thirteen-year-old daughter to a wheelchair had been dismissed as a mere stomachache.

Professional athlete Kyle Warner had been public about his injury on social media, especially about what he was doing to recover as he tackled the task at hand like any professional athlete would. And Doug Cameron, who had been encouraged by his employer to have the Johnson & Johnson vaccine, went from being an active rancher to being confined to a wheelchair after inflammation in his spinal column ruptured, severing his spine. There was Cody Flint, an injured agricultural pilot and young family man who could no longer fly; Shaun Barcavage, a research nurse practitioner who was now on disability; and Suzanna Newell, a triathlete whose vaccine had left her dependent on a walker for months, and with severe autoimmune stiff person syndrome. Kellai Strodriguez had been struck down by unpredictable tremors, making it almost impossible to hold her young baby. They were all willing to go to Washington on behalf of those who couldn't or wouldn't.

Senator Johnson's office sent Dr. Joel Wallskog to the group, suggesting that he would be a good candidate for speaking. He was an orthopedic surgeon whose injury meant that he could no longer perform surgery. And a sympathetic media contact introduced Bri to Ernest Ramirez, a grieving father who had lost his sixteen-year-old son suddenly to myocarditis five days after vaccination. Ernest had been planning to travel to DC from Texas, to find some way to share his son's story. Now he didn't have to do it alone.

Ernest had already worked out how to get himself to the event, with his buddies accompanying him for moral support. Bri was responsible for getting the injured there, a task that turned out to be far more complicated than she'd expected. And expensive.

There was no way she would have expected everyone to pay their own way to the event—they were all injured, not working, and financing their own healthcare. Going to Washington involved travel and accommodation costs for sixteen people; and after the roundtable, Bri wanted to organize an awareness-raising event on the steps of the Supreme Court, to make the most of them all being in the same place at the same time. Everything would cost money.

Bri put $22,000 of it on her credit card. Brian was *not* happy. But Bri was convinced that the exposure the event would bring to the vaccine-injured would be worth it, and anyway, they had started to attract some wealthy and influential individuals who claimed to be in the injured's corner—they could put their money where their mouths were and sponsor the travel, if nothing else.

To Bri's surprise, most of the people she approached declined. *Wait . . . You said you supported us!* She was shocked to discover that some of the people who had been loudest on social media about the "poor vaccine victims" went quiet in their private conversations when asked to *actually* do something that could help. One offered $2,500 toward the costs of the event but only if he could speak at it. Bri declined—the event couldn't be used for any other agenda. It was an opportunity to highlight the plight of the vaccine-injured *only*.

The Children's Health Defense, chaired by Robert F. Kennedy Jr., offered $2,500 with no strings attached; no recognition was required. CHD was a non-profit organization founded in 2007, with a mission to end childhood health epidemics by eliminating toxic exposure. The organization had been questioning childhood vaccines and calling for further investigation into their safety profiles for years; they were very concerned about the administering of Covid vaccines to children and keen to support the roundtable.

The CHD donation helped, but much more was needed. Bri wondered if she could use her story to generate funding, so she did an interview with independent niche outlet *LifeSiteNews* about her own injury experience and asked for financial support to get the injured to the Washington event. The site featured donation links periodically throughout the article they ran, and over $26,000 was donated on that crowdfunding platform alone. In total they managed to raise $37,000 to pay for everybody's travel and accommodation, *and* the rally afterwards. *People do care!*

On top of the fundraising, there were the logistics to arrange—renting a bus to bring them all from the hotel to the Capitol Campus, coordinating with Capitol security, helping the injured prepare their speeches according to strict timings set by Senator Johnson, and issuing invitations to media, filmmakers, musicians, lawyers; representatives from the FDA, the CDC, and the NIH; and every single senator.

Finally, they were going to be heard.

* * *

They first met in person in the lobby of the Holiday Inn, where they were all staying, but a bystander would have assumed they were friends who had known each other forever. Despite the solemnity of the occasion, they greeted each other with huge smiles and warm hugs, then sat laughing and joking together, exuding love. Vaccine-injured communities throughout the country—and throughout the world—would find the same joy when in one another's company. There was a unique bond between them; a bond that perhaps only those who have been in combat together would understand.

The scientists were surprised. This rather jolly entourage was not what they were expecting. And again, this was happening the world over, as the feisty and compassionate community battled their way through preconceptions and prejudice from those who believed them as well as those who didn't. The vaccine-injured *always* surprised onlookers with their ability to smile despite their circumstances; there were no victims here. Instead, there were unexpectedly cheerful warriors.

At the back of the lobby stood a group of tattooed men wearing black leather, eyeing the exuberance on the other side of the room. Bri was in preschool teacher mode, organizing everybody like she was running a field trip, until she spotted Ernest Ramirez and his motorcycle friends. She walked straight up to him, arms outstretched, pulling him in for a big hug. Ernest was taken aback; here was this skinny white lady treating him like an old friend. He had been grieving his son for what felt like so long now, that he had forgotten what love felt like.

Aside from Steph and Maddie, it was the first time for Bri to meet everyone off-screen. There had been countless meetings beforehand, especially between Bri and Dr. Linda Wastila. Linda was well-known and highly respected in her field, and in the weeks leading up to the roundtable, Linda had found herself awake in the middle of the night, terrified of what her participation might do to her career. She knew of others who had lost academic positions after speaking out against mandates, and this would be the first time she would be going public about them herself.

There had been plenty of other professionals who had been asked to speak; professionals who had quietly expressed their concerns about the vaccines or vaccination policies to Linda and Peter, but were uncomfortable expressing them publicly. Nonetheless, Linda and Peter had managed to gather an impressive group of expert speakers. There was Lieutenant Colonel Theresa Long, MD, a physician in the United States Air Force who had been seeing vaccine injuries among the pilots; retired GP and former president of the UK Royal College of General Practitioners, Iona Heath, who had been honored by the Queen; risk assessment expert from MIT, Retsef Levi; civil rights and vaccine attorney Aaron Siri; molecular biologist with experience in mRNA technology, Dr. Aditi Bhargava; pediatric and rheumatology specialist Dr. Patrick Whelan whose publications had been cited over 69,000 times; and pharmaceutical harm expert, Dr. David Healy. All were willing to put their careers on the line, in support of the vaccine-injured.

It was the first time for the injured—and bereaved—to feel that their stories mattered.

The next day, they were all much more subdued as they waited in a very long line to go through Capitol security, conserving their energy and trying to mentally prepare for what had been scheduled as a three-hour roundtable in—coincidentally, given one of their supporters—the historic Kennedy Room. Microphones and name cards were placed along the tables, with Senator Johnson front and center, flanked by Dr. Doshi and Bri. At the back were two rows of chairs with more name cards— Dr. Anthony Fauci, Dr. Janet Woodcock, and Dr. Rochelle Walensky. None of the representatives of the federal health or government departments turned up. They didn't even send a representative.

Neither did any senators or mainstream media, although Bri's own senator, Mike Lee, sent a team of staff who stayed for the entire event.

But representatives from independent media like *The High Wire* and *The Epoch Times* did turn up. The entire event was livestreamed through their uncensored platforms, but everyone had been disappointed to learn that there had been an unprecedented last-minute rule change by the Senate committee that meant that the testimonies they would be making would not be put on record in the Senate. It was a decision that had never before been made. And it meant that there wouldn't be an official record to show the roundtable even happened.

The event went on for four hours—an hour longer than originally planned. The scientists shared their expert knowledge, research, and concerns about the Covid vaccines and associated mandates; they provided compelling data, much of which had never been discussed in any kind of public forum. However, the personal testimonies of the injured were what everybody remembered from the day; the testimonies were a reminder that behind the science, *real* people's lives were being affected, every single day. Including Christmas Day, which Ernest shared he would be spending at his teenage son's grave.

The entire room was in tears. There were tears from the injured as they bravely told their stories to an audience that gave them the space they needed to do so; there were tears from the scientists when faced with the human devastation behind the data; and there were tears from the media at the realization that the people in front of them weren't just a story— they were a warning of a world that was losing its sense of humanity.

Overwhelmed at actually being heard, and also relieved that the roundtable was over, the injured answered the reporters' questions, then headed outside for the next part of the day—the rally on the steps of the Supreme Court that overlooks Capitol Hill.

The injured took their positions on the steps in front of the Supreme Court in pouring rain; some with canes, some with walkers, some in wheelchairs. Even the most physically able had to be seated because of the challenges that standing for any length of time would bring. Several hundred supporters joined them, carrying homemade signs and banners. It was here that their loved ones could join them too—the senate's strict rules after January 6 meant that they hadn't been allowed to attend the roundtable event to offer physical or emotional support.

Political musicians Hi-Rez & Jimmy Levy sang their powerful lyrics against forced vaccination and the growing loss of personal and medical autonomy in the United States. And the microphone was passed to the vaccine-injured who hadn't participated in the roundtable, as well as those who had. Bri had been keen to ensure that any injured who wanted a voice had one on those steps.

Senator Johnson stood on the side of the crowd, declining offers of umbrellas, instead ensuring his staff brought umbrellas for the injured. He quietly observed the people who were making such an impact on him that he was now considering running for reelection. He was the only politician present at the roundtable, and also the only politician to stand outside in solidarity with the injured, in the rain.

Cody Flint—the injured pilot—had gone to the voting floor and taken the opportunity to talk directly with some of the senators he saw there. Every one he spoke to said that they had no idea that the roundtable event was happening, despite them and their staff being invited to it well before-hand by Johnson's office and Bri's team.

Everybody else returned to the hotel, a little despondent because of the lack of mainstream media and political interest in their efforts to be heard, but excited to know that the event was going viral online—links to Senator Johnson's congressional website and *The HighWire* were being shared all over the world. YouTube would remove the recording within days, but for now, the injured delighted in reading the supportive comments below the video.

And again, they delighted in simply being with each other; their injured family.

The next day they met individually with their own representatives, handing over professional binders prepared by Cody. There were pages and pages of information about VAERS reports, failing vaccine compensation policies, and personal testimonies. Bri met with Senator Lee and had an in-depth conversation about compensation reform that prompted him to put forward a bill that would ultimately fail to gain support by the rest of the Senate, but she appreciated his effort.

Bri's spirits were high upon her return to Utah, but her body felt utterly broken. Still getting used to her new limitations, Bri had pushed herself far beyond what her body was capable of now, and there were repercussions that led to her being confined to bed again. But now, she had reporters to talk to from her bed; the roundtable had sparked media interest.

One of those interested media representatives contacted the clinical trial company about Bri's medical expenses. Soon afterwards, the clinical trial company offered the Dressens $1,243.30 as a "full and final settlement," along with a request to relinquish AstraZeneca from all responsibility for any future expenses. The offer felt rushed; her name wasn't even spelled correctly.

Bri was lying in a hospital at the time the offer came through, receiving $3,500 worth of an IVIG infusion; AstraZeneca's offer didn't even cover half of that, and the consent form had clearly stated that all medical expenses, and other costs, would be reimbursed.

As Bri's lawyers from Aaron Siri's law firm, Siri & Glimstad, went through the two consent forms that Bri had signed—one she'd had fully explained to her on the day she got the shot, and the amended one she'd signed after she got sick—one of the attorneys noticed that the one she'd felt pressured to sign contained some significant changes, and referred to conditions *exactly* like the ones that Bri had been suffering with for over a year.

In the amended form, the information about "Serious Reactions" was longer . . . much longer, and included reactions that had not been mentioned *at all* in the first consent form. There were warnings about blood vessels and blood flow, potential organ and tissue damage, neurological disorders, spinal cord issues, blurred vision, weakness, numbness, tingling, trouble walking, and bowel/bladder issues. There was reference to a trial participant that had developed serious nervous system symptoms, the cause of which was at that time being evaluated, and trial participants were told to urgently seek medical help for any of those symptoms, because "They may cause substantial disability, and some can be fatal if not treated promptly."

The amended form was created on November 6—the day Bri reported her reaction. *They knew all along. They KNEW and didn't tell me.*

Bri had spent months in and out of hospitals with all the symptoms that had been added to the consent form. If only someone had taken the time to explain the changes to her when she'd signed it. If only somebody had said, "That's not normal, but we are aware of this happening." If only her doctors had been given a clue to look for these debilitating conditions, then maybe somebody would have taken her more seriously? And if the conditions had been identified and treated earlier, who knew what damage could have been prevented? Who knew what further damage may have been done to her in all this time? *How can my life mean so little to them? My life? $1,200?*

All the injured who risked their health and their reputations by speaking at the roundtable wondered the same thing as the mainstream media and government agencies stayed silent—*how can we mean so little to everyone?*

Not everyone. They meant everything to each other.

The gathering in the Kennedy Room and outside the Supreme Court had created a commitment to one another that surprised the injured themselves. It was like their injuries had ignited a fierce protection toward each other, akin to a familial bond. The connection brought about a sense of purpose to compassionate, driven, high-achieving individuals whose purpose in life had been unexpectedly and dramatically cut off. No longer able to focus on their previous personal or professional pursuits, they looked at their fellow injured, were moved by the injustice of the situation, and were compelled to be part of the solution. Coming together on November 2, 2021 changed many of them forever, especially Dr. Joel Wallskog.

Joel had developed transverse myelitis following a single shot of Moderna, after which he was given a medical exemption for further shots and told to take at least two months off work. Dedicated to his career as an orthopedic surgeon and not being someone who tended to take time off, Joel had returned to work after just two weeks. On his second day back, his legs collapsed from underneath him, and he went completely numb from his waist down. That was in January 2021, and he hadn't returned to work for the rest of the year. He would likely never perform surgery again. His much-loved career of almost twenty years was over.

Joel had taken all the appropriate steps to report his reaction, submitting his details to VAERS, and contacting the CDC directly. Like the other injured, he expected vaccine injury specialists to contact him not only to offer a treatment protocol, but to gain information that could be of use to the vaccine manufacturers and other scientists, researchers, and injured. He was shocked when that didn't happen.

He had spent much of 2021 battling intense feelings of abandonment, exacerbated by his workers' compensation claim being denied. He was unable to enjoy his previously active life with his family of five children and seven dogs, and every day spent lying on the sofa may have eased his physical pain, but constantly fed the emotional pain he felt at no longer living the life he had worked so hard toward.

When Senator Johnson had told Joel about the roundtable plans, and introduced him to Bri, he agreed to share his story at the event, but wasn't prepared for how his meeting of the other injured would have such a profound impact on him. On the steps of the Supreme Court—despite his own challenges that impacted his legs, and an old-fashioned gentleman at heart—Joel held an umbrella over the heads of the other injured as they took the microphone, one by one, and told stories of their own suffering, their own pain, their own abandonment. He stood behind them, listening to every word and realized that this was far bigger than him and his own pain. He returned to his hotel room, called his wife, and told her that he had found his purpose in life again. He knew what he had to do. He had to help the other vaccine-injured.

Using the small amount of leftover funds from the roundtable event, Joel formalized the group that Bri and the others had been trying to put together. They called it React19.

They were going global.

CHAPTER 10

GOING GLOBAL

The injured and the experts who spoke at Senator Johnson's first round-table event on November 2, 2021, had expected both the science and the suffering to be all over the mainstream media in the days and weeks that followed. They were disappointed to find that not to be the case.

However, there was a *little* bit of coverage. A few news reports appeared, always starting with how many people the virus had killed and how many the vaccine had saved, always saying that the vaccine was "safe and effective," always talking about how rare reactions to it were, and always ending with a statement declaring that vaccines were the most effective way to beat Covid.

But in between all of that, there was *some* telling of the stories of the humans that were suffering, and even some questioning of the processes involved in producing this vaccine, if not all vaccines in general.

Bri was sitting next to Peter Doshi at the roundtable when the *BMJ*—which he edited—published the results of an investigation the journal had conducted into claims made by Pfizer whistleblower Brook Jackson. Brook had been a regional director at one of the company's clinical trial sites and had expressed serious concerns she'd had with the trial at her site.

The day after the roundtable, a different story ran on *LifeSiteNews*—the site through which Bri had shared her story to raise funds for the vaccine-injured to travel to Washington. The site published a letter that vaccine-injury lawyer Aaron Siri (also at the roundtable) had sent to the

FDA, CDC, and HHS (US Department of Health & Human Services) a week earlier. The letter informed the government departments about serious harms that had occurred post-vaccination and were not being addressed. It included signed testimonies from eleven confirmed vaccine-injured—all of whom were themselves healthcare workers—and asked for acknowledgment of, research into, and treatment for the damage that was being done. Danice and Mary had included their testimonies.

The news wasn't all encouraging though. Two days after the round-table, the US Department of Labor's Occupational Safety and Health Administration (OSHA) issued a nationwide order to all companies with over one hundred employees, instructing them to mandate those employees to get the Covid vaccine, or wear a face covering and get regularly tested. To those already injured and struggling to keep their jobs, the order for mandatory vaccination was devastating. *Our suffering really doesn't matter at all.*

Any questions about the vaccines or mention of vaccine injuries had to be quashed. The vaccine-injured were not good PR for a vaccine the government was mandating.

NBC News ran a lengthy story about vaccine misinformation and "anti-vaccine propaganda," mentioning new policies that social media platforms were implementing in order to control the sharing of vaccine-related information. Several people were named as being key to what was called an "anti-vaccine" movement, including Del Bigtree, host and producer of *The HighWire*—one of the few platforms to broadcast the roundtable event. The news story sent a clear message that anyone speaking about what *might* be an adverse reaction to a vaccine, was the same as someone who was either cautiously questioning or confidently against vaccines; the fact that the former needed diagnosis and treatment didn't seem to matter. They were all just "anti-vaccine," a term that was used twenty-two times in the article, which included several paragraphs on Maddie, and questioned whether the cause of her condition was actually related to the vaccine at all.

Clips of the roundtable event were going viral online, and Maddie quickly became the poster child for the vaccine-injured. She *loathed* the publicity—they all did. None of them were speaking out for attention.

They wanted nothing more than to focus on their own healing, but instead reluctantly found themselves to be public figures for the injured community. Speaking out felt like the right and responsible thing to do—a duty to humanity—but there was a price to pay. Others would later emerge within the community who would embrace the attention their new illness would bring them on a national and international stage, but most of the advocates were extremely uncomfortable with it. The attention wasn't always pleasant.

Kyle in particular, was hounded by a group of moms who were very much in favor of vaccination, and they saw the vaccine-injured as a threat to their beliefs. Kyle already had a very public platform from his life before injury, when he had worked hard to establish his credibility not just as a professional athlete and businessperson, but also as a decent human being. A genuinely good person, Kyle was well-liked in the biking world and was used to being treated with kindness and respect. The hatred that came his way after he did the roundtable was shocking to him, and he had no idea how to deal with it. Blocking the perpetrators on his social media account didn't help; they just made new accounts and harassed him from those. He was exhausted enough dealing with his illness; now he had online abuse to deal with as well, and he ended up having PTSD because of the relentless attacks from people like this group of women. He had never experienced that level of hate directed at him before, and he struggled to process it. He just wanted to get back to biking again, but like all of them, connecting with the other injured in Washington had changed him.

The scientists and doctors who had spoken out were changed too, and not only by the words spoken by the injured who sat alongside them. They'd known they'd be risking their careers and now they had to face the reality that entailed. They were called in to meetings with their superiors and questioned about what they thought they were doing speaking at such an event, and reminded that they represented the organization for which they worked. Some of them were reassigned to other projects or moved into less significant roles. All got the message, loud and clear. Speaking about vaccine injuries was unacceptable.

Bri struggled with the guilt she felt as she watched the pain among her new friends as they were criticized and controlled, many of whom she had personally encouraged to speak out. Some of the experts had had cold feet as the event had drawn closer, and she had sent them the injureds' testimonies that they'd spent hours working on, which of course recalibrated the experts' focus. All of them—injured or expert—had been so *incredibly* brave. She felt they should be celebrated, not silenced.

Bri herself received an extraordinary amount of attention after the roundtable. Her social media was flooded with messages and comments from people all over the world, many of whom were themselves injured and in desperate need of someone—*anyone*—who might be able to help them. In Bri, they saw a very articulate and determined woman sitting next to a senator, talking about the *exact* same things they were dealing with, and they reached out to her. In a matter of days, she had 3,000 friend requests waiting on Facebook.

It was impossible to keep up with all the attention. Bri spent hours wading through it all while simultaneously trying to protect herself against the angry vitriol from those who claimed to be "pro-vaccine." There were also the vicious messages from people on the other side of the debate to deal with, and she tried to process the cold-heartedness of those who thought she deserved her illness for being stupid enough to get the vaccine in the first place. It felt like the vaccine-injured were hated by everyone. But Bri couldn't allow the onslaught of unkindness to affect her. She focused on scrolling through the sea of messages, trying to identify who was actually injured and in need of help. *If I'd done this before, then Heidi would still be alive.*

In between all the advocacy and attention, Bri was still trying to keep herself alive. Viewers may have seen an articulate and determined version of Bri at the roundtable, but the effort it took to be *that* Bri was beyond what anybody watching could have imagined. The doctor who her sister had arranged had managed to get Bri's insurance company to cover the IVIG treatments that she was having twice a month, and the treatments *were* helping, but one year after she'd had that single dose of AstraZeneca, her life was still unrecognizable. She was *still* in constant pain; *still* shaking; *still* vibrating; and *still* experiencing electric shock sensations throughout her body. And she was still losing weight—she could keep food down now, but eating had not yet returned to being the joy it once was. It felt like time was running out.

While Bri's public demeanor was authentic and brutally honest as she spoke about the realities of being vaccine-injured, her private life depended on her finding some way to *pretend* that everything was fine. Instead of having fun *with* her children at a park, or camping, or biking, or swimming, Bri spent her home life observing them as they went about their own lives, trying to put the thoughts of anger, resentment, and sadness to the back of her mind. *Yes honey, I did see that cartwheel! Yes, I can help you with your homework from here. Yes, I can sit up enough to do some painting with you.*

She knew that this was a world away from where she was at the beginning of 2021, when she was lying in the bedroom upstairs, unable to even bear the sound of her children's voices, the touch of their little hands, or the feeling of Hannah snuggling up close, but this new "half-presence" in their lives almost made her feel worse. Somehow it reinforced just how much her life had changed. She was no longer removed from their lives; she was an observer, which in some ways felt even more painful. Now she could *see* the life she could no longer participate in as it played out right in front of her. And it broke her heart. *Is this going to be the rest of my life?*

At least now, she was able to look her kids in the eyes, and smile. She was able to pretend that even though she was horizontal, she felt fine. She was able to love them again, if nothing else. She was still here.

And while she was still here, Bri was going to make sure her suffering wasn't in vain. She was going to do everything she could to make sure others who were suffering knew they weren't alone. If the NIH, the CDC, or the FDA weren't going to step up for the vaccine-injured, then they would step up for themselves. React19 would step up.

Bri, Danice, Sheryl, and Kristi had set up a nonprofit organization to provide financial, physical, and emotional support to the injured community. They needed to be formally recognized if they ever had any hope of being taken seriously by more senators, scientists, or sponsors. Danice did not want to be in the limelight and, after the attention that Bri had received following the November roundtable, Bri really didn't blame her. But Bri hadn't wanted to be the sole figurehead. She had developed a close connection with Joel, and he had been very knowledgeable and helpful as they planned the registering of a 501(c)(3). He eventually gave in to Bri's persuasion and agreed to be listed as "co-founder" with her.

They intended for React19 to be a very pro-science organization both in terms of the practical information it could share with the injured community, and in terms of specifically encouraging scientists to either work with or research them.

To everyone's surprise, Kristi's health had greatly improved after she'd tested positive for Covid; it was like the infection had corrected her immune system. But she couldn't walk away from her new friends who were still struggling, so she agreed to be the organization's secretary. Sheryl had managed to return to school to complete the course she had been doing when she'd become injured. She was very careful about not taking on too much so didn't become a board member, but stayed within the new organization to offer support wherever she could.

With the number of injured who were only just finding them, and the new injured that would be coming in after the mandates, they needed more board members to share the workload. They turned to those who had spoken so eloquently at the roundtable, thinking they would be appropriate representatives of the injured community. Kyle, Shaun, Suzanna, Ernest, and Cody all agreed to join the board. While there were obviously differences in opinion, they shared the same desire to provide compassionate support to the vaccine-injured; and to be the friend that they had all wished they'd had when they'd first become sick themselves.

While React19 was established to support the American injured, it would ultimately become a hub for injured people based all over the world, and a model for vaccine-injury organizations in almost twenty different countries. Each of these countries were facing so many of the same issues, on top of different problems that were specific to that country's political or social system. As Bri learned more and more about adverse reactions to other vaccines, she realized that the vaccine-injury support groups of the past had faced the same problems that the Covid vaccine-injury support groups were facing now—medical gaslighting, media censorship, lack of treatment pathways, and government refusal to acknowledge their existence. *This has been going on for decades right across the globe.*

What the vaccine-injury organizations of the past didn't have, however, was a unified, global voice. Never before had any vaccine been administered to so many people at the same time; the *New York Times* reported that five and a half *billion* people (almost three quarters of the planet's entire population) had received at least one dose of a Covid vaccine. This had been a global vaccination program, and it would have global repercussions unlike anything that had ever been seen or dealt with before. There would be people suffering with adverse reactions all over the world, and the world didn't seem to be *at all* prepared for it. Bri reasoned that the injured needed to come together, advocate together, and ultimately heal together. But more importantly, right now—at the end of 2021—they needed to learn together. Japan had dealt with HPV vaccine injury in 2013, and both the UK and Sweden had dealt with swine flu vaccine-induced narcolepsy in 2011. There would be experts around the world who had dealt with smaller versions of what the Covid vaccine-injured were dealing with.

Bri reached out to the new international friends she had been making since she'd gone public earlier that year—many of whom were running their own support groups—and started bringing each of them in on a group chat that would become known as the React19 International Coalition (or ReactGlobal). As the chat developed and more international injured leaders were added to it, the group developed their own declaration—a formal document clearly stating their unified mission and goals, agreed upon and signed by them all. And they found a connection with each other that mirrored the connections they had developed within their own communities—they weren't just advocates, or support group leaders; they were friends.

Never before had the vaccine-injured unified across the globe in their compassion and commitment toward each other.

Mia Wolff in Germany was one of Bri's first international vaccine-injured friends. After a single dose of Moderna, Mia had experienced a reaction very similar to Bri's, and was keen to use her scientific background not only to try to get to the bottom of what had happened to her, but also to be able to support the vaccine-injured that were connecting with each other in her country. Unlike other countries, the German media wasn't censoring conversations about vaccine injury. Caution was used, but because freedom of speech was valued and protected, the German press had relatively more freedom than other countries. Also, the pharmaceutical industry was restricted in the lobbying and marketing they were permitted to do in Germany, so questions about vaccines were more acceptable. Vaccine-injury was a slightly less taboo topic in Germany than it was in other countries.

Germany had even established a post-vaccine clinic in the University Hospital Marburg, and in the first six months of 2022, treated 250 people and had a waiting list of 3,000. The country's Health Minister, Karl Lauterbach, publicly committed to helping people with long-term consequences of Covid vaccination, pledging to ensure vaccine injury would be recognized more quickly, and planning research on "Post-Vac Syndrome."

Germany was also one of the countries where the vaccine-injured could go for apheresis—a kind of "blood washing" procedure that was becoming popular among Long Covid, vaccine-injured, and ME sufferers. With vaccine-injury stigma being far less of a barrier in Germany, it was possible for sympathetic and interested scientists and physicians to focus on solutions and treatments, which is what Mia was working hard on.

However, there was a general consensus in Germany that theirs was the only country where vaccine injuries existed. Mia needed support in educating the scientists, government, and media so that they understood that adverse reactions to the Covid vaccine were affecting people all over the world. She was a valuable asset to the global group, and the other members were able to provide her with information she could use to show what was happening in other countries.

Bri also worked closely with UK-based Charlet Crichton, who she roped in to help manage the global chat group. Charlet had set up UKCVFamily around the same time that React19 had been established—the UK had a very different healthcare system than the US, and the injured over there were struggling to navigate their way through it. It wasn't like in America where you could independently make an appointment with a specialist or go get your own testing done; the British were completely beholden to their GPs for *anything*. So if the GP wasn't sympathetic to vaccine-injury, or misdiagnosed someone because "vaccine-injury" didn't exist in NHS guidelines, the UK injured were completely stuck.

Charlet had been debilitated by her second AstraZeneca vaccine; although in hindsight she'd realized that some of the "odd" health problems she'd had after her first that had been dismissed as early menopause were actually vaccine-related. Menopause was a common misdiagnosis among the predominantly female vaccine-injured across the globe—even women in their twenties were being put on HRT, which more often than not made their symptoms worse.

Like Bri, Charlet was no longer able to work in the business she had created. Before vaccination, Charlet had led an incredibly active life; even her work as a sports therapist was physically demanding. Now, she was rarely upright. She would be propped up in her bed surrounded by cushions when she joined calls, yet was always able to smile. Bri found such a softness in Charlet; she was warm and sweet, and she just wanted to fix everything for everyone. She'd been volunteering in a vaccination center when she'd been offered a vaccine from what was left over that day. Like Bri, she just thought she was doing the right thing.

Now Charlet was doing the right thing in a different way. She was leading a support group that would ultimately become the first ever registered charity for the vaccine-injured in the UK. Alongside Charlet was a formidable team of equally motivated and compassionate people dedicating their time to supporting their fellow injured. Despite the gentle demeanor, Charlet was a force to be reckoned with.

Bri loved all the Brits she connected with—she loved their sense of humor and the way they seemed determined to make each other laugh despite their suffering. Most of the injured in the UK had been injured by AstraZeneca, which was still being celebrated as a great British success story. In the summer of 2021, the Oxford-based scientists behind the development of the AstraZeneca vaccine had been honored by the Queen, something that did not sit well with the UK vaccine-injured who had been begging the pharmaceutical company for help for months, only to be ignored. Two of the scientists had even written a book about their roles; neither Bri nor Charlet were even able to read a book, but the injured that could noted that it was sadly lacking in any humanity toward the vaccine-injured and instead read more like a defense of the product they had created. *If those scientists had spent their time helping us rather than writing that book, maybe some of us would be able to read again.*

In Australia, there was Rado Faletič, the "rocket man." Rado was a scientist in hypersonic technology who, like Charlet, had experienced odd symptoms after his first vaccine, but a more severe reaction after the second. He had been vaccinated in the fall of 2021, much later than many others in Australia, because he had educated himself about the technology behind what he understood to be a new kind of pharmaceutical product. He waited to see if there would be any cause for concern, but the Australian media was full of stories urging everyone to get vaccinated with no mention of adverse reactions. Telling himself that if there *had* been any problems, the government would have alerted the public to them by then, he went ahead to get his shots.

Parts of Australia had been strict in its exclusion of the unvaccinated to normal life. Only fully vaccinated people were permitted access to certain services or in some cases even to go to work. Unvaccinated teachers were not allowed in school, but if they had a sympathetic headteacher, their jobs would be held for them while they went on unpaid leave. When they returned to work after mandates were lifted, they were served a letter of formal reprimand for putting their students' lives at risk, and had their salaries docked as punishment, despite having risked their own health to keep schools open for the vulnerable in the early stages of the pandemic before vaccines.

The restrictions led to a toxic culture and traumatization of the unvaccinated from which it would be difficult for the country to recover.

Rado had spent much of his professional life connecting researchers, and had great networking skills, but was struggling to get scientists or the TGA (Therapeutic Goods Administration) to take his adverse reaction seriously. His frustration led to him co-founding **CO**VERSE, an Australian charity modelled on React19.

In New Zealand, Bri connected with Edoardo Galli and Anna Hodgkinson. Edoardo was an Italian PhD student who was living with the long-term impact of an adverse reaction to a single dose of Pfizer. Frustrated by the lack of knowledge among his doctors, and the lack of any kind of pharmaceutical intervention available, Edoardo was following an entirely natural approach to his own healing, which he was keen to share with other injured. Bri loved talking to Edoardo and his fiancé—they were a young, cute couple that reminded her of herself and Brian when they'd first met.

Anna Hodgkinson was the founder of SilentNoMore NZ, a support group established after the difficulties she'd also had with the New Zealand healthcare system, but she wasn't injured herself—her teenage daughter was. Bri always felt like it was one thing to be living the pain of vaccine-injury yourself, but to be the *mother* of an injured child—like Steph and Anna—was a whole other kind of pain. Anna was very public; not just trying to get help for her daughter but for all the other New Zealand injured too. She had even managed to get an appointment with the prime minister, to which she'd invited Bri, but the meeting was canceled at the last minute.

Julie Bertone was based in Belgium and running a French-speaking support group, as well as pushing for research among French-speaking scientists. France also had severe restrictions placed on the unvaccinated, with mandates being imposed since the fall of 2021. Three thousand healthcare staff had been suspended for not getting the Covid vaccine, leading to protests and staff shortages across the country. France did not feel like a safe space for the vaccine-injured to talk about their struggles.

Julie had been severely affected by a similar reaction to Bri's and had also found that food had a big impact on her symptoms. Despite her suffering, she exuded warmth and kindness, speaking softly during group calls. Like Charlet, there was an underestimated strength beneath her gentle manner.

Bri brought in another Pfizer trial participant to ReactGlobal. Augusto Germán Roux was based in Buenos Aires and had participated in a trial in Argentina. He became unwell after his first vaccine but continued with the trial, with his health worsening after the second jab. He collapsed soon after and was misdiagnosed with having Covid. Like Bri and Maddie, he was abandoned by the trial clinic, by the pharmaceutical company, and by his government. In the absence of any healthcare, he had started doing his own research into vaccination procedures, and was coming to the conclusion that the adverse reactions in *all* the trials had been hidden or misrepresented.

Augusto was a powerful and influential lawyer in Argentina and was in the process of very quietly gathering all the information he needed to file a lawsuit. Bri was happy to help him however she could, and connected him to the growing group.

Stevan Mihajlovic was the youngest of the group, still in his twenties. Stevan was a software developer from Serbia and had had a severe reaction to a single jab that got even worse after he caught Covid. He had lost everything, including his girlfriend, and was struggling to get doctors to believe him; however, like Edoardo, he had been doing his own research and was keen to share what was working for him. Stevan was based in Switzerland, where assisted suicide was offered as a way of dealing with chronic health conditions. In the months to come, *many* of the vaccine-injured community would consider Switzerland as an option. Some of them went ahead with that option.

The React19 International Coalition rapidly grew to almost fifty members; all representatives of their own country's vaccine-injured communities, and all doing their best to ensure that the people who looked to them for support didn't go looking to Switzerland for a solution.

It was an extraordinary level of responsibility.

Bri saw the very same pressures and problems among her new international friends that she was experiencing herself. They were all public figures within their own vaccine-injured communities if not further afield, they were all trying to balance their own health limitations with the practicalities that running a group entailed, and they all had similar items on their to-do lists—usually involving attempts to communicate with scientists, politicians, or media. They were all fighting an uphill battle to get thousands and thousands of very sick people the help—and simple human kindness—they needed. And many of them were doing that from their own sickbeds.

Bri found tremendous power within the global group. Whatever she perceived to be her weaknesses as an advocate, there was always somebody else for whom that was their strength. She could post a message on the group chat about something she'd remembered just before going to sleep, perhaps asking if there was anybody willing to talk to a very science-based publication about a specific kind of symptom, and by the time she'd woken up there would be countless messages with names of people the group leaders would suggest, along with personal introductions. Together they would brainstorm ideas for awareness-raising campaigns, come up with ways to get around the ongoing censorship, share the results of polls they'd run within their own communities, and jump in to defend each other if one of them was getting a particularly high level of online abuse.

Most importantly, any of them that were contacted by another injured who had happened to find them online, now had a global network to draw upon. The first question Bri would ask was, "Where are you?" and she would send them straight over to the person heading up that country's efforts so that they were no longer alone.

As Bri became more internationally known for her adverse reaction and advocacy, she was no longer the person she'd been a year earlier, searching online to find the support she needed. Now she *was* the support. She'd found such special friendships among the first injured she'd met—Danice, Sheryl, Kristi, and Candace—and the American advocacy efforts of 2021 would never have happened without them. Now, out of necessity, those friendships shifted into more businesslike relationships. They were no longer a bunch of friends on a Zoom moaning about their symptoms; they were now board members of an official organization, with agendas, minutes, accounts, audits, and public reporting responsibilities. They were also targets for individuals and organizations that had their own agendas in mind.

Everything that accompanied dealing with others' injuries made it difficult for Bri to find time to deal with her own.

Bri had never been very good at putting her own needs first. She had never learned how to create and impose boundaries. She had never learned to say "no" to anyone who wanted her to do anything. When her siblings had told her to jump off a cliff, she'd jumped, ignoring the instincts that told her what she was about to do was very, very dangerous.

The international leadership role was about to force Bri to learn how to put her own needs first, how to set and stick to boundaries, and how to say "no." She had to connect with her instinct and intuition so that she would become able to figure out who *genuinely* wanted to help the vaccine-injured, and who would ultimately cause more trauma not just for her but for the entire vaccine-injured community.

Because not everybody was quite the ally they claimed to be.

PART THREE

THE INJUSTICE

CHAPTER 11

FINDING ALLIES

For most of 2021, Bri, Danice, and Sheryl had been reassured that Dr. Nath and his team at the NIH were on the vaccine-injured's side. As the injured community had rapidly grown around them, they had repeatedly told some very desperate people that the NIH were helping. They had said that the NIH would be publishing their research any day now, after which even more of them could get medical help.

But there had been no sign of the research paper that Dr. Nath had promised. Despite their suspicions that the NIH may have been dissuading journalists from covering the vaccine-injured, Bri held on to hope and kept sending journalists to Dr. Nath; she had plenty of emails as proof that the NIH was working with the injured and had been until the point where the study was halted. After the roundtable event, independent news outlet *The Epoch Times* was keen to be a very supportive media ally to the injured, and wanted to write about the NIH research, but Dr. Nath told journalist Zack Steiber that the NIH wasn't doing any research on vaccine injuries. So Bri sent *Science* magazine—the news outlet for the AAAS (American Academy for the Advancement of Science)—to the NIH. Finally, after being confronted with what was regarded a reputable news outlet, and a year after the research started, Dr. Nath admitted that they had in fact been studying a small group of people who had become sick after their Covid vaccines, but downplayed any possible link and urged readers to be cautious about drawing any conclusions. Bri was shocked to read his quotes in the article, remembering how supportive Dr. Nath had been right from the very first time she met him during that online meeting where she could barely hold up her head. *He believed us! He didn't question us! I thought he wanted to learn about WHY we were injured, not IF we were injured!*

The NIH study was finally published as a preprint four months after the *Science* magazine article publicized the controversial study. It was deeply disappointing, based on only twenty-three patients, whereas they had been sent more than a hundred. The paper concluded ". . . a variety of neuropathic symptoms may manifest after SARS-CoV-2 vaccinations and in some patients might be an immune-mediated process." It also stated that further studies were needed.

Unsurprisingly, the NIH's follow-up survey for fifty vaccine-injured participants languished with only thirty people nationwide enrolling. *Do they think we are too complicated? Are we incurable? Have we got some kind of weird syndrome that they don't want people to know about? Are they waiting for us to just die and disappear?*

It had been a huge blow to Bri when Dr. Nath had distanced himself from them all, and it had become apparent that he was not the ally she had thought. While Olivia and Candace had said early on that the NIH was a waste of time, Bri had trusted the government. She had trusted everyone she was supposed to trust: the clinical trial company, AstraZeneca, the media, and the government. All of the institutions in which she had invested her faith had let her down, and her trust had been completely shattered. Many of the vaccine-injured throughout the world felt the same—not only were their bodies failing them, but the entities that ordinary, everyday people trusted to be there in the event of any kind of life-shattering event were failing them too.

With no support, and absolutely nowhere to turn, it made the vaccine-injured communities extremely vulnerable to individuals and organizations who had their own agendas.

When React19 representatives started getting invited to speak at events, they accepted the invitations with a significant amount of trepidation that wasn't just because of possible repercussions for their health. Their trust had been fundamentally broken. They'd learned that the people they had *thought* were allies weren't at all. And they'd learned that there were people in the world who considered the vaccine-injured to be enemies. They knew what it felt like to be hated and didn't know if they had the strength to absorb more of that. They didn't want to be reliving their own personal trauma time and time again unless it was going to directly benefit their community. They didn't want to align with something that could be distracting from their central message—getting support for the vaccine-injured. They were willing to pay a price, but it had to be worth it. They had to get something out of it in return.

So when they were offered time to speak at the "Defeat the Mandates" rally in Washington in January 2022, they cautiously decided they *would* take that risk, but they would use the opportunity to get something that the vaccine-injured really needed . . . money for medical bills.

Joel and Kyle addressed the thousands of attendees, as well as thousands more who were watching live, and spoke for twelve minutes about what was happening to the Covid vaccine-injured in America. They shocked even the most informed attendee, and the donations started pouring in.

They raised over $100,000 in less than fifteen minutes.

They felt listened to, and they felt *loved*. There was no hate. The crowds had greeted them with open arms, not only allowing their own tears to flow but giving the injured the permission they needed to allow theirs to flow too. Their stories were safe here.

The rally in Washington was an incredible platform for the vaccine-injured to be given, for which Bri and React19 were extremely grateful. It connected them with a public that cared, even if there were questions over the motives of some of the people creating opportunities to reach that public.

Over time, Bri had a sense that among certain groups, it was kind of cool to be seen to be associated with the vaccine-injured. But for many "allies," that was the extent of the association. It was just for appearances. And this was happening across the globe with other injured advocates— events that were often promoted as if they were grounded within the ethics of equality, freedom, and respect; yet the event itself would have a clear hierarchy, with certain doctors and scientists at the very top. Some had become household names among those who were very much aware of what was happening in 2020 and the years that followed. These doctors and scientists were treated like celebrities, given access to VIP rooms, speaking fees, and some even had their own security staff. Everyone assumed that they were working with and treating the vaccine-injured, but as Bri, Joel, and Kyle tried to network at these kinds of events in the early months of 2022, they felt invisible. Some of the influential people didn't even stick around to hear the injured speak but instead returned to the VIP space, seemingly focused on who was the next most influential person they could be seen to be speaking to.

The few "Covid celebrities" who *were* willing to give the vaccine-injured a platform were also very clever and motivated by a good business sense. Giving the vaccine-injured a platform to be seen usually came with strings attached. Getting a vaccine-injured person to talk about the most terrifying day of their lives, and ongoing disability because of that day, was guaranteed to draw a lot of web traffic or podcast viewers in certain circles. Even more if the host could get them to cry on camera. A show with a high-profile vaccine-injured person like Bri would bring in a lot of revenue, but not to her or any other injured . . . the revenue would go to the host. The most memorable was a podcast by a medical non-profit organization that claimed to have a protocol to cure the injured as well as those with Long Covid. Bri was a guest on it, and that episode raised $37,000 for the host. Not one cent was donated to React19. But anyone watching would have assumed that the host was a hero, stepping up to support the injured.

This was the Wild West Covid "freedom movement" that pulled the vaccine-injured advocates back and forth, to suit their own needs rather than the needs of the injured movement.

Some mainstream media organizations finally did step up in 2022, thanks to an extensive campaign by React19. Formalizing their organization did pave the way to be taken more seriously by some journalists, editors, publishers, and producers. Stories started to appear in the news, not just about the adverse reactions that some people were suffering from, but also about the difficulties the injured were having in being recognized and treated by doctors, and even the censorship they were being subjected to by online media. Mainstream news articles were appearing in other countries too, usually focusing on someone who had been injured or bereaved, and always beginning and ending the story with a statement about how safe and effective the vaccines were. But after a complete block on any and all information related to vaccine harms, the coverage was a step in the right direction.

Health-specific publications started calling for more research, sometimes into very specific symptoms such as tinnitus (something many of the injured suffered with for the first time since getting vaccinated) and sometimes into the complex combination of symptoms in general. The lack of medical and scientific knowledge was being acknowledged, very slowly.

As 2022 went on, there was an increase in stories questioning the efficacy of the vaccines, or whether the boosters were even necessary; both topics that React19 as an organization stayed away from, but nonetheless, such coverage did contribute to a social climate where more and more injured felt increasingly comfortable to talk about what had happened to them, and less afraid of the stigma associated with vaccine injury.

Some very select media were becoming *somewhat* of an ally, but the injured leadership had to take a huge leap of faith before speaking to journalists. Some journalists claimed to be interested in writing about adverse reactions, but all they really wanted to do was ask some questions in an effort to paint the people talking about them as "crazy conspiracy theorist anti-vaxxers"—much like the journalists did with Senator Johnson. But it was hard to defame the injured advocates because they were genuinely calm and collected as they spoke about their personal experiences rather than posing any theories. Any journalists hoping to expose people they assumed were just *claiming* to be injured would lose interest when it would become clear that that was not the case at all. These were normal people, who had done what their government asked them to, and had been abandoned by everyone.

There were a few very special journalists who were genuinely sympathetic. They would be in tears when learning about the level of suffering within this hidden sector of society. They became overwhelmed at the stories they were told; unable to get the vaccine-injured out of their minds or their hearts. In the UK, former ITV and BSKYB News Executive Mark Sharman was so moved by the vaccine-injured people he got to know, that he came out of retirement and produced the hour-long documentary, *Safe and Effective: A Second Opinion*. Mainstream media wasn't interested, so it was released on YouTube in October 2022, to be removed by the platform within hours of being mentioned in a parliamentary debate, and having almost reached a million views. It was the first material entirely created by a non-injured person that sensitively, accurately, and respectfully portrayed the injured, and without any other agenda.

Back in the US, as the advocates struggled to get more mainstream media involved in their plight, they were slowly working out who among the alternative media they could *really* trust.

Daniel O'Connor was the owner of *TrialSiteNews*, an online, uncensored news portal about clinical trials. The site's content was produced by experts within the clinical trial industry—physicians, lawyers, and regulators—and had been launched by Daniel in 2018. The business had transparency as its core mission, and less than two years after its inception was playing a key role in the effort to counteract the singular narrative being promoted by the government and mainstream media. They were publishing stories about the vaccines almost every day, and following the formalization of the support group, were specifically covering React19's activities on a weekly basis.

Daniel had some very interesting information about the AstraZeneca trial. In the summer of 2020, he had heard of claims that some companies in Southern California had struck a deal with the pharmaceutical company and were about to mandate their illegal immigrant employees to participate in the AstraZeneca trials. Daniel had been in the process of launching an investigation into the claims when the trial had been abruptly put on hold, and he never knew why. Once he met Bri, and she explained what the trial clinic had told her about the possible adverse reactions in the UK trial and the need to pause for investigation, Daniel finally understood.

In Bri's experience, many of the journalists who wanted to cover the vaccine-injured had a genuine desire to help, but when they realized just how complicated the situation was, how sick people really were, how much of a thick skin was really required, and how much abuse they were at risk of receiving, even the most well-intentioned would quietly step away, exhausted and overwhelmed with what it really meant to be an ally to the injured. Not Daniel. He frequently made it clear to Bri that he was in it for the long haul.

Jared St. Clair was another ally with a huge platform that he was happy to offer to the injured. Jared grew up working in his parents' nutrition store. After his parents passed away, Jared continued the family business, selling natural products relating to health and wellness. He believed very much in the body's ability to heal itself and had kept the doors of Vitality Nutrition open during lockdowns. His employees went mask-free, and he would not be taking the Covid vaccine himself. He had been the subject of much abuse for his beliefs, at one point being confronted by a masked woman in his store. She had been wearing a shirt that said "vaccines saves lives," pulled down her mask, and spat directly into Jared's face, saying, "I hope you die of Covid."

It would have been understandable if Jared had struggled to connect with someone in a mask who had taken a Covid vaccine, but he connected with Bri as soon as he met her; injured, frail, weak, and wearing a mask. Not only did he connect with her, but he wanted to help take her pain away. He became a very special person to the Dressens and to the injured community.

As Jared's friendship with the React19 community grew, he didn't once ask Bri to appear on his podcast, and instead always seemed to be focused on what *he* could do to help her and the community around her. He knew about the FDA processes and was an expert on natural heath products. He knew that there were ways to maximize the body's ability to heal.

Jared wanted to use his seventeen-year experience as a radio host to help elevate the stories of the injured, and set up a regular podcast for React19. The *Dearly Discarded* podcast, consisting of a series of interviews with the injured, was launched in April 2022, with Jared's kindness and compassion evident in every conversation. Like many other healthy people living normal lives, for Jared becoming close to the injured community was a deeply moving experience. He frequently found himself in tears after the shows; after having listened to each and every word of one person after the next and the multiple ways in which their lives had been destroyed.

It was one thing to know that vaccine injuries existed; it was quite another to spend an hour connecting face-to-face with someone who was living that existence every single day.

Another ally who was frequently shocked at the extent of the suffering she witnessed among the vaccine-injured was Dr. Suzanne Gazda, an integrative neurologist based in Texas. While some medical professionals were prominent on the Covid celebrity circuit, some more than others were actually doing anything of value directly with the injured. Dr. Gazda was one of them, and she became the expert to whom React19 would send its most complicated cases. She was constantly following the latest research into Long Covid, and sending Bri important papers related to severe immunological or neurological complications. She knew that understanding of Covid-related illness was evolving, didn't claim to have any magic cure, and valued the patient's perspective as well as encouraged them to take ownership of the elements of their own healing they could control themselves. Like the practitioners around the world who *were* helping the injured—physically, medically, emotionally, or spiritually—she was doing so quietly in the background without seeking out the attention and accolades that others seemed to be more focused on. Dr. Gazda was playing a vital role in the treatment of those with vaccine injuries and Long Covid.

Sections of the Long Covid community itself were also allies to the injured, although the relationship between the two groups was not without conflict. Some people who had been diagnosed with Long Covid before the vaccine was available believed that if it had been made available sooner, then they wouldn't have become ill. Some that had been bereaved by the virus assigned a certain blame to the vaccine-injured—in some cases directly accusing their vaccine-injured friend as being responsible for the death of a loved one who had refused to get vaccinated. And likewise, some of the vaccine-injured harbored resentment toward those with Long Covid—the Long Covid patients were used as reasons to push everyone to get vaccinated. Long Covid clinics were established throughout the country, but the vaccine-injured, as opposed to the Covid-injured, were for the most part denied access to them.

Many of the injured had been misdiagnosed with Long Covid despite having never contracted Covid. Some of the Long Covid groups wouldn't allow any discussions about symptoms post-vaccination, and even banned anyone who tried to mention the topic in their support groups, whereas some of the groups encouraged an open dialogue. And it was there—in the middle of all the anger, resentment, blame, silencing, and *fear*—where seriously ill and very vulnerable people remembered their sense of humanity and reached out to each other to share love and support. When it came down to it, they had *all* had their lives taken away from them. Did it really matter what had done that?

Finding common ground and coming together in a country and a culture where differences—in politics, in religion, in race, in vaccination status—had been emphasized for so long, was the most powerful reminder that we are all people, all capable of compassion, and all capable of deciding *not* to take a side.

One of Bri's strongest allies would turn out to be a very public figure with whom, for her whole life, she had always assumed she would have no common ground whatsoever. Someone who, unbeknownst to her, would be announcing his run for presidency specifically under a campaign around finding common ground and coming together—Robert F. Kennedy Jr.

She was reluctant to get on a call with him at first. He was just another crazy, right? He was one of the Disinformation Dozen who the government had issued warnings about. He was an anti-vaxxer conspiracy theorist with a drug history and questionable relationships with women, wasn't he? Kennedy was the last person she and React19 needed to be aligning with; it would be way too risky.

But she wasn't comfortable with the word "anti-vaxxer" anymore. She wasn't sure when a "conspiracy theorist" just became a "theorist." Some of the people providing her marginalized community with unwavering support would probably fit very well within that Disinformation Dozen. And her experiences with Senator Johnson had shown the senator to be nothing like how he'd been presented in the mainstream media. *Maybe I need to just make up my own mind?*

Kennedy wanted to talk about censorship. He had been deplatformed himself for almost twenty years, was knowledgeable about adverse reactions to vaccines, and he was interested in the Covid vaccine-injured community's experience of being silenced. Bri agreed to meet with him and prepared by gathering together all the proof she had of their community being censored. Just like when Bri had met Johnson, she was surprised to find Kennedy to be quite unlike *anything* she had ever read in the media about him. Like Johnson, Kennedy was genuinely kind, extremely knowledgeable, and fully aware of the stories that circulated about him. He had learned ways of dealing with the online abuse that Bri was just beginning to learn herself.

Kennedy kept in touch with Bri by text and invited her to a censorship protest he was organizing outside Facebook headquarters. She accepted the invitation and despite being back in a wheelchair again, she made it to the event in May 2022, determined to confront Mark Zuckerberg. But she was at another low point emotionally—all her advocacy efforts felt like they were resulting in such little progress, yet were costing her personally in any kind of real recovery she might be able to make if she was able to just focus on herself. She was doing everything she could to fight for the injured but felt like she was fighting a losing battle in more ways than one.

"How do we win?" she asked, feeling very dejected.

Kennedy told Bri of his fight to protect the environment for years, with only the occasional win here and there. Despite everything he had done, the environment was still getting polluted by major corporations and the environment would continue to get worse; it was a daily battle for him, but it was one that he would never give up on.

"I will fight to protect the environment until my dying breath." He told Bri that she had to be prepared to do the same, and asked her if she was ready for that fight.

Then he shared something his father had taught him . . .

"In the civil war there was a general directing his troops to battle. They were outnumbered and the troops weren't sure how to move forward. They asked their general, 'What if we lose?' The general had replied, 'It's not about winning or losing. We are all going to die one day, either with liberty or without liberty. Some will die on the battleground today. We are all ultimately headed to that same fate, and we get to decide how we get there. Do we die fighting, or do we die without fighting at all?'"

Kennedy gave Bri the pep talk she needed without asking for anything in return, and has continued to do so ever since. Along with Senator Johnson, Bri had found another very unexpected ally behind the image that the media had given her to believe in all those years. Whenever she visited Washington, DC, with any of the other React19 leadership or members, she knew she could count on Johnson if something unexpected would arise. During one visit with wheelchair-bound Andre Cherry and his own senator, the wheelchair broke. At the end of the meeting, the senator and his staff simply left the room, saying that they were late for their next meeting. Not knowing what to do, Andre's new friends *carried* him and the broken wheelchair to Johnson's office for help. Johnson's staff immediately jumped out of their seats to see what they could do. Senator Ron Johnson—and his entire team—continued to treat the vaccine-injured with kindness and respect. They still do.

Bri was gradually finding genuine allies not just for the injured movement but also for herself personally among the political, legal, medical, and media worlds. And there was another very unexpected ally that entered her life in 2022—an ally from an entirely different world: the world of entertainment.

Jessica Sutta had been one of the Pussycat Dolls, an all-female singing and dancing pop group. The band had achieved worldwide fame in the mid-nineties, but disbanded by the end of the following decade, and Jessica was pursuing a solo career. She was the only former band member to reach the top of an American chart, not once but four times. She'd delayed having the Covid vaccine until the birth of her first child. Her reaction started three days after a single shot of Moderna, when she started having multiple sclerosis-type symptoms, but nobody associated them with the vaccine, including herself. The second shot completely debilitated the formerly fit and healthy pop star, and a neurologist told her that they were aware of about fifty other patients with similar reactions to the Covid vaccine, but refused to discuss it at a follow-up appointment.

Like so many others, Jessica was lost with a new illness that nobody seemed to be able to help her with. She lay awake one night, her ribs burning, and her spine full of stabbing pains, as she searched for anything that could possibly help. She came across Bri, Joel, and Ernest speaking at the roundtable. Shocked, Jessica reached out to the newly formed React19 website, nervous about giving her real name. She had to talk to Bri.

Bri's heart broke as she heard Jessica talk about how all her dreams of raising her baby boy were completely shattered—she couldn't even pick him up. Jessica was devastated at the thought of her son being robbed of the vibrant, active, life-loving woman that his mother used to be. And of herself never being able to dance again, something that brought her so much peace and balance during difficult times.

Bri understood. She felt the very same pain every single day. All Jessica had been offered so far was one form of benzodiazepine or another, and the world of music and fame that Jessica occupied led to her to becoming vehemently against taking anything that could be addictive. She had seen enough damage that addictive substances had done.

Jessica had access to the best of the best medical professionals, but she could not find a single one that knew how to help her.

So Bri helped her, gently guiding her on what she needed to do to calm down her immune system, to tweak her already healthy diet, and to adjust her lifestyle so that it would accommodate her new limitations. Over time, the two women found a deep connection, and as Jessica became more able to manage her own symptoms, she became an unexpected source of support for Bri in return.

Jessica had spent her youth and adult life in the professional music industry, working with the likes of Britney Spears as well as other household names. She knew what it was like to be in the glare of the spotlight. She knew what it was like to be taken advantage of. She knew what it was like to live life unsure whether somebody genuinely wanted to connect, or whether somebody just wanted to satisfy their own narcissistic needs by positioning themselves close to her. Over her years as a well-known pop star, Jessica had learned how to set and stick to boundaries, was familiar with status climbers, and had developed a pretty good radar at spotting unscrupulous people. Bri was still struggling with all of these things.

Jessica saw this clearly, even though Bri was still blind to much of it. Jessica wanted to protect her so became Bri's ally and confidante. And eventually joined the React19 board.

Along with the original group of injured—Danice, Sheryl, Kristi, and Candace—and the new injured friends Bri found in Steph, Joel, and Kyle, Jessica played a vital role in helping Bri learn how to put her own health first.

After a significant improvement toward the end of 2021 when she was having regular IVIG, Bri's health deteriorated in the spring of 2022 when she and her entire family tested positive for Covid for the first time. The kids just slept a lot but apart from that, it was no different from any other virus they had had in the past. Hannah seemed to have it worse than Cooper, and sought only the security of her mom, curling up next to her at every opportunity. By that point, Bri had learned about early interventions that some people were having success with, despite there being very little positive mention of such interventions in the media or by the government. The family managed their symptoms with antihistamines, lots of fluids, and really clean eating. With their trust in the media completely nonexistent, the Dressens felt comfortable using pharmaceuticals that had been given very bad press, and everybody recovered within weeks.

A bout of the virus did, however, seem to ramp up all of Bri's vaccine-related symptoms, and ultimately was the catalyst for her return to a wheelchair just before she went to help Kennedy confront Zuckerberg at the Palo Alto campus. She was trying to balance her health as well as her React19 responsibilities, and it was tough. The toughest point was when the people who had supported her during the early months—her cousin who shaved her legs, the other who wrote the well-intentioned letter, and other family members as well as some friends—all became sick with Covid, and Bri was too sick to be able to support them in the way they had supported her.

Despite the official guidance that all children were to be vaccinated, the NIH had privately advised the Dressens not to vaccinate their kids. At no point did neither Bri nor Brian wish they had. It took nine months for Bri to get back to her pre-Covid baseline of health, but she never wished she'd gone back for a second shot. She still struggled with so much regret that she'd gotten the first, and wished that she had been willing to take her chances with Covid itself. *At least then the spike would have been filtered through my body, instead of being MADE by my own body.*

The way that Bri and all her new friends were being treated by the vaccine manufacturers, the healthcare system, the pharmaceutical industry, the CDC, the FDA, and the NIH had led the Dressens to seriously reevaluate everything they thought they knew about vaccines. Ordinary Americans were being completely abandoned to muddle their way through one scientific paper after another, in the hope of finding any kind of clue that could give them their lives back.

But the clues were there to be found.

CHAPTER 12

FINDING THE SCIENCE

"Follow the science" was repeated by governments and media around the world before the Covid vaccines hit the market. Questions about them were met with, "follow the science," rather than an open discussion about what that science actually said, not to mention who—or what—was behind the science. Where did the science come from? Who was funding it? Who was approving it? On what basis? Very few of the vaccine-injured had had any kind of conversation with their own physicians about whether they should get the vaccine or not, and if they did, most medical practitioners repeated what they had been told—the Covid vaccine was safe and effective. The science said so.

But how much of the science had the physicians themselves actually read? In all the chaos of 2020, the last thing any busy family doctors had time for was to study scientific papers about either Covid or the vaccine. They didn't have the time for lengthy conversations with patients about the science behind government authorization or approval of the new products. They trusted the government health agencies as much as the general population did. If the FDA said to get vaccinated, then that was enough for them.

The people for whom FDA authorization/approval was *not* enough; the people who wanted more information or discussion before deciding whether to get vaccinated or not; the people who actively sought out scientific information about the clinical trials—they were labeled "anti-vaxxers."

America became a country where educating oneself about science showed stupidity, and accepting the science without question showed intelligence.

But any scientist would agree that science is there to be tested, challenged, and discussed over and over again. Science is a field that continually evolves; very much dependent on individuals with curious minds who are motivated enough to pursue answers to the questions and theories they feel free to explore. True innovators welcome the questioning of their work and relish the opportunity to expand on scientific knowledge.

But from 2020, "following the science" meant accepting that the *only* way of managing Covid-related illness was to get vaccinated. It meant accepting that other pharmaceutical products—products that had been safely used for decades—would be useless against Covid. And non-pharmaceutical interventions? They were irrelevant. What kind of idiot would think that nutrition or lifestyle had anything to do with helping the human body cope with illness? No, vaccines were the only way out of the pandemic. Science said so.

And if science said so; if science said that these new Covid vaccines were the *only* way out of the pandemic, then the law permitted the FDA to give emergency use authorization (EUA) to them, even if they hadn't gone through the same clinical trial procedures that other pharmaceutical products had in the past. The public could rest assured that these new products would be closely monitored for any adverse reactions, which would be swiftly dealt with . . . by science.

But Bri knew differently. And the thousands of vaccine-injured that had found React19 knew differently too. Nobody was monitoring them. Nobody was dealing with them. Science wasn't interested in them at all.

Who *was* supposed to be monitoring them? Who *was* supposed to be researching adverse reactions to the Covid vaccine, the largest vaccination rollout in history? How was it decided that there was something concerning about any vaccine? And who decided whether the public should know that there was a concern?

The United States' Vaccine Adverse Event Reporting System (VAERS) was cited by physicians, politicians, and government health agencies as being the source of any potential safety signal with the Covid vaccines.

Most countries had their own similar systems: Canada had the Canadian Adverse Events Following Immunization Surveillance System (CAEFISS), and the UK had the Yellow Card System, which allowed the reporting of adverse reactions to all pharmaceuticals, not just vaccines. Throughout the world, the vaccine-injured were being told that the vaccines were safe and effective, and if there was a problem then their country's reporting system would have flagged up those problems.

Except the reporting systems weren't working. And hadn't worked for years.

Patients didn't necessarily know about the reporting systems, and physicians didn't always use them, despite being obliged to in certain situations. A Harvard Medical School three-year study conducted a decade before Covid concluded that despite adverse reactions to pharmaceuticals occurring in one in four patients, less than 0.3 percent were reported. For vaccine adverse reactions, the study found that it was less than 1 percent. The study went on to explain that low reporting rates were mainly due to doctors being too busy; the process of reporting was complex and time-consuming. Another study, published in the scientific journal Vaccine in 2013, found that 37 percent of healthcare practitioners had come across an adverse reaction yet only 17 percent of them had reported it.

For 2021 and 2022, most healthcare professionals operating in what was likely the most challenging time of their entire careers—how many of them really would have had the time and inclination to report a possible adverse reaction to a vaccine?

The raw data of the VAERS reports was all supposed to be public, published online, and made available for download by year, starting from 1990, when there had been over 2,000 reports. The number of reports had steadily risen in the thirty years that followed, to almost 50,000 reports in 2020.

In 2021, VAERS logged over 750,000 reports of adverse reactions. The Covid vaccine reports outnumbered all reports, from all other vaccines combined, since the reporting system began.

But if only 17 percent of practitioners who had recognized vaccine adverse events were reporting them, and less than 1 percent being reported at all, what were the real figures? By the end of 2021, almost 250 million

Americans had had at least one dose of a Covid vaccine. Twenty million more would go on to have a first dose. Bri wondered how many of them were out there, suffering alone, with no idea what *could* be behind their complicated new conditions.

The VAERS website was clear about the coincidental nature of what may be assumed to be a reaction to a vaccination—that blood clot that appeared twenty-four hours after a shot may well have happened regardless. But few of the vaccine-injured were dealing with just *one* symptom. Most of them were dealing with an onslaught of ten, twenty, thirty *life-altering* symptoms within days of being vaccinated. Surely such reports warranted scientific investigation?

But not all the reports were being followed up. Not even by phone. While the US government had found the funds to employ countless staff to operate the contact tracing system designed to minimize Covid infections, it didn't seem that contacting the injured was a priority. The vaccine-injured were reporting their reactions wherever they were supposed to, under the assumption that a team of scientists would be in touch within days to get as much information as possible not just with the intention of treating this person who had done what their government had asked them to, but to protect others in the future.

At the end of 2022, React19 discovered that 1 in 3 of the VAERS reports weren't even being processed properly. The group conducted their own audit of over one hundred reports that had been filed by members who had been concerned about difficulties they were having in accessing their reports. The audit found over a third of the reports were either not being made public or had been deleted altogether. The reports that were visible inaccurately represented the medically recognized status of the injuries—permanent disabilities were medically recorded for 53 percent of cases, yet VAERS only recorded 23 percent. VAERS had even recorded a report of death as merely "hospitalized."

The data was supposed to give science the opportunity to investigate safety signals, but the data wasn't accurate, and the science wasn't happening. While the general public assumed that adverse events were being monitored by the government agencies, the FDA was only concerned with safety monitoring *before* vaccines got authorized, not after.

Bri had been devastated when the NIH had stopped their study. All the injured involved in the study had. For much of 2021, it had given them their only hope of having any scientific research about their conditions published. Nonetheless, Danice in particular had persisted in trying to get noticed by other scientists and had managed to spark the interest and sympathies of medical researchers at some prestigious institutions.

Stanford University was evaluating autonomic dysfunction after Covid infection and was willing to look at cases where the condition occurred post-vaccination. Johns Hopkins University had been running all of the neuropathy tissue assays for the NIH study. Scientists at Mt. Sinai Hospital and Cornell University were working on papers to help define the cluster of symptoms. Mayo Clinic had submitted a paper on a case series of three patients suffering severe neurologic and autonomic dysfunction after Covid vaccination, the University of Vermont was studying one hundred small fiber neuropathy cases, and Yale University researchers were studying immunophenotyping (a kind of "fingerprinting" of immune cells) in both Long Covid and vaccine-injured groups.

These world-renowned scientists were following the exact same methods and guidelines that had become the norm for Long Covid research. The only difference was that the patient groups they were studying—or incorporating into wider studies—were vaccine-injured.

None of the studies were accepted by the scientific community.

The studies were either stopped, or the resulting papers were rejected, sometimes by multiple journals. Bri was told by one scientist that they had never experienced multiple rejections in their entire career. Yale Immunology—while applauded for their research into Long Covid—managed to get their vaccination-related research published on their fourth attempt, only to have it withdrawn at a later date with the explanation that publishing such research was dangerous and irresponsible; reviewers specifically noted that recognizing Danice and Bri as coauthors was a factor in the eleventh hour rejection. Could patients not be involved in research? Had science become a field where only safe and responsible research was considered appropriate? Who decided what counted as safe and responsible research?

The injured advocates spent much of 2022 relentlessly but respect-fully emailing the people at the top of the FDA: Robert Califf, Janet Woodcock, Peter Marks, CDC's newly appointed Rochelle Walensky, Jay Butler, Daniel Jernigan. They were all made aware of the challenges the vaccine-injured were facing in having their conditions recognized or researched. They were clearly told of the obstacles that were being placed in the way of recovery for the people whose lives had been destroyed. And those very same people whose lives were destroyed were willingly offering whatever information they could, wanting to be active partners with the FDA in a collective search for solutions.

The officials often responded, and their responses didn't deny what was happening to this determined and increasingly desperate group of con-cerned citizens. Peter Marks often shared his own perspective . . .

"Nobody is denying adverse events can occur."

"Nobody is denying the symptoms of what's going on here."

"No-one is denying there are [sic] vaccine injury here. Nobody is deny-ing that."

"No-one is denying there are reports of neuropathy or potential of vaccine injury here."

"Nobody is saying they're not real."

"We have never said you guys don't exist."

Bri's response to Dr. Marks expressed the frustration felt by all the injured. "We know we exist. We know you know that we exist. But the public doesn't think that you guys know we exist."

The FDA seemed interested in gathering information, but not so inter-ested in doing anything *with* that information. It was so simple. *Just tell the public that you know about us!*

But the FDA stayed publicly silent, all the while maintaining private communication with the Covid vaccine-injured.

Woodcock, Marks, and co. rarely had anything helpful to offer, and as the year went on, desperation among the injured was evident in their ongoing communication; healthcare professionals who had dedicated their lives to caring for others and who had truly believed in the healthcare system were now trying to use their medical knowledge and communica-tion skills to speak up for others. But they ultimately found their faith in the country they lived in—and the individuals and institutions at the head of it—completely and utterly shattered.

Dr. Nath had made it clear to Bri that vaccine monitoring was out of the NIH's remit, telling her that the vaccine manufacturers—profit-making companies—should be the ones taking charge, and that it wasn't the government's responsibility to "pick up after them."

The government wasn't willing to pick up after the vaccine manufacturers yet had created a system whereby the vaccine manufacturers didn't have to directly pick up after themselves. That system went back to 1986 when the National Childhood Vaccine Injury Act (NCVIA) was passed. The act mandated the reporting of adverse reactions; established a committee to review scientific papers on adverse reactions; set up a National Vaccine Program Office to coordinate the CDC, NIH, and FDA's activities; and required an information statement be given to the patient before every vaccination. It also removed any financial liability from the manufacturers due to vaccine injury claims or death, and instead established the National Vaccine Injury Compensation Program (VICP).

The VICP was a jury-free system of assessing vaccine-injury claims which, if awarded in the case of a recognized injury, would be considered "no fault." Any alleged unrecognized (or new) injuries would not be considered "no fault" and the petitioner would have a very high burden of proof. If successful, VICP claims were paid for by a fund supported by a tax the manufacturers paid per vaccine. VICP claimants often ended up waiting *years* for their cases to go to court.

However, the Covid vaccine injuries didn't come under the VICP. They came under a different program—the Countermeasures Injury Compensation Program, or CICP.

The CICP applied to anyone injured as a result of interventions during a public health crisis that had necessitated invocation of the PREP Act . . . such as a vaccine during a pandemic.

The PREP Act (Public Readiness and Emergency Preparedness Act) was established in 2005, giving pharmaceutical companies immunity from state or federal litigation in the event of a public health emergency, including pandemics. It also provided protection in cases of willful misconduct and gave anyone involved in the administering of a pharmaceutical product complete immunity from litigation.

Whereas the VICP was funded by a tax paid for by the pharmaceutical industry, the CICP was funded by the American taxpayer.

Injuries resulting from the Covid vaccine came under the CICP because the PREP Act was invoked on February 4, 2020, giving Covid vaccine manufacturers no reason to be accountable for them. Nobody could sue them. Anyone claiming to be vaccine-injured could apply to the CICP, but to be successful in a CICP claim, there had to be reliable and credible proof that the Covid vaccine could cause whatever symptoms the claimant was suffering from; reliable and credible proof that had been recognized in scientific journals . . . but the journals weren't publishing any of it.

The vaccine manufacturers themselves were not liable for any damage caused to the recipients of their products. The companies had no reason to conduct or even support any scientific research that could lead to diagnosis or management of any injuries. They had no reason to share information about their own product that could lead to the development of treatment pathways for people that product had harmed.

Manufacturers of the Covid vaccines had no reason to even speak to anyone suffering, even if there was absolutely no question of the suffering being due to the administration of their own product.

As was evident in how AstraZeneca was treating Bri. The pharmaceutical company didn't even ask for Bri's medical records until March 2022—sixteen months after her reaction—and then only for the purpose of assessing her claim for reimbursement for medical expenses, not for the purpose of investigating her injuries. After emailing back and forth until September, AstraZeneca finally said that they had everything they needed. That was the first and last time Bri ever heard directly from them.

It was up to the vaccine-injured themselves to find the science to follow in order to develop their own treatment pathways. It wasn't like there wasn't *any* science out there; there had been studies on adverse reactions to other vaccines in the past. You just had to dig long and hard enough to find them.

There was scientific research into adverse reactions to diphtheria, polio, flu, MMR (measles, mumps, rubella), and HPV (human papillomavirus) vaccines spanning well over a century. Adverse reactions to vaccines weren't new and had in many cases been swiftly responded to with investigation, awareness campaigns, policy development, and safer practices being put into place. What was dubbed the "first modern medical disaster" involved a contaminated diphtheria vaccine in 1901, and a year later there was a similar incident with a smallpox vaccine. The disasters led to the creation of the Center for Biologics Evaluation and Research, the very organization of which Peter Marks was now the head.

Another contaminated diphtheria vaccine incident occurred almost thirty years later, this time in Australia, and resulted in the deaths of twelve children. The vaccination program had immediately been stopped not just in Australia but in other countries too, despite diphtheria being the leading cause of childhood death at the time. Within five months of the deaths there had been a royal commission, multiple papers in scientific journals, and a government commitment to providing financial compensation and the covering of ongoing medical expenses for the survivors.

In the 1950s, production problems with a polio vaccine were found to have caused the disease itself in 40,000 of the 200,000 children who had received it. Ten children died, two hundred were left paralyzed, and the rollout was stopped within a month of the adverse reactions. Another polio vaccine was found to unexpectedly reactivate in the gut, and also *cause* the disease it was meant to prevent. This was especially interesting to the Covid vaccine-injured, with many of them having symptoms not dissimilar to Long Covid.

In the nineties, an increase in meningitis cases led to scientific papers on an association with some MMR vaccines, depending on the strain of mumps contained in the vaccine. Meningitis was a known complication following the actual disease and was not expected to occur following vaccination. The vaccines containing that strain were withdrawn in the UK. Meningitis was one of the conditions that some of the Covid vaccine-injured were getting a diagnosis for.

In Europe, over 1,300 people developed narcolepsy after receiving a flu vaccination in 2009. Multiple scientific papers existed spanning decades of research into how the vaccine could have caused this ongoing neurological condition, with some symptoms mirroring those of the Covid vaccine-injured.

A few years later, Japan withdrew its recommendations for the HPV vaccine after 2,000 teenage girls reported the development of a complex syndrome—a syndrome not unlike the combination of POTS, CFS, fibromyalgia, and cognitive difficulties that the Covid vaccine-injured were dealing with. Also, like many of the Covid vaccine-injured, many of the teenage girls had been treated as if they had a psychiatric disorder. Scientists from other countries also had research published into similar adverse events following HPV vaccination.

Bri and Brian were stunned by the scientific research that was out there. *How can there be scientifically documented incidents spanning well over a century, yet people still pretend that adverse reactions to vaccines don't exist?*

They had already learned that the Covid vaccine didn't work like the vaccines that had preceded it. Smallpox, polio, MMR, HPV, flu vaccines all worked by injecting a very small amount of a weakened or inactive version of a virus, which then prompted the body's immune system to develop antibodies. If at a later date, the person was exposed to the actual virus, the antibodies would go to work. The Covid vaccines didn't work this way. Instead, they involved injecting the person with "instructions" so that the body would actually make a part of the virus itself. This part was called the spike protein.

Unlike Covid vaccine injuries ("Long Vax") scientific research into Long *Covid* was extensive, and often discussed the implications of the spike protein part of the virus. This proved to be useful to finding clues that might help the vaccine-injured. The spike protein was found to be neurotoxic—it could cause long-lasting or permanent damage to cells, especially those involved in brain signaling. Symptoms of neurotoxicity were similar to symptoms of poisoning, mold or chemical exposure, or excessive drug use (prescribed or recreational).

Long Covid scientific research was saying that the spike protein spread throughout the body, reaching bones, tissue, and organs, as well as circulating throughout the bloodstream. The pharmaceutical companies were saying that the vaccine—or instructions telling the body to create the spike protein—was less risky than the virus because it would just stay in the arm.

But a Japanese study commissioned by Pfizer said something quite different, and found that the lipid nanoparticles traveled throughout the entire body—the spleen, brain, reproductive organs, adrenal glands. The thought terrified Bri, who lived every day feeling like her body was poisoning itself. *Are our whole bodies producing this neurotoxic spike protein?*

A renowned AIDS medical researcher contacted Bri, asking for a sample of her blood. His team was discovering that the spike protein didn't seem to be clearing in those with ongoing symptoms post-vaccination. As Bri was one of the first people in America to get a Covid vaccine, she had been living with symptoms for longer than anyone else the scientists could find. Could the spike protein be the cause?

The scientists' theory proved to be correct. Not only did Bri have vaccine-derived spike protein in her blood, the scientist told her that she had more spike protein than anyone they had tested so far. But the spike protein they were finding in the vaccine-injured study participants was different from the virus-derived spike protein. This was a new, "mutant" spike.

The researcher asked Bri to introduce them to the FDA so they could share this potentially vital discovery. Perhaps a certain sector of the population were unable to clear the spike protein from their systems? Perhaps some people were making a different type of spike protein? Perhaps that explained why some people had zero symptoms post-vaccination—some people's bodies may have been clearing it without any problems at all.

The FDA didn't respond to Bri's repeated attempts to connect them with the researcher, and she never heard from the researcher again.

But she didn't give up. Neither did her new friends. Within the vaccine-injured community there were plenty of scientists and medical professionals who were dedicated to getting to the bottom of what was happening to their bodies. They felt the weight of the increasing numbers of lives that needed saving as their community grew bigger and bigger, all across the globe. Unable to return to their previously busy lives, some of these highly educated and scientifically minded individuals—including many who were themselves physicians—spent hours and hours scouring research papers from their beds and sofas as they searched for clues.

Sometimes the clues came from papers written by a very unexpected person.

Dr. Nath had contributed to a paper published as early as March 2021 in the *Annals of Neurology* journal, discussing possible neurological complications following Covid vaccination. Bri's attention was drawn to the suggestion that if someone developed such a reaction, then it would seem "prudent to avoid that vaccination in the future." She recalled how the support groups were full of severely injured people who experienced neurological adverse events to their first dose and were then pressured by their doctors to go on to get more shots.

Later that same year, Dr. Nath had coauthored another paper on neurological disease and the Covid vaccine, this time in the prestigious *Neurology* journal, and giving hints as to possible reasons that could indicate appropriate treatment pathways, "They are thought to be immune mediated and early recognition and treatment with immunomodulatory therapies might be warranted . . ."

Dr. Nath mentioned clotting issues post-vaccine which, thanks to the efforts of a team of expert hematologists in the UK, had been recognized in a number of scientific papers and mentioned in the UK media. VITT (vaccine-induced thrombocytopenia and thrombosis) was a new clotting disorder specifically related to an immune response to the Covid vaccine, had no cure, and had led to fatalities. The condition had led to the AstraZeneca vaccine being paused in Europe, and ultimately not being administered to the those under thirty in certain countries. Patient information leaflets had been updated to include mention of the condition, and NHS guidelines had been created for treatment pathways that included mental health support.

VITT had garnered the kind of scientific recognition in the UK that one would have expected for all of the serious adverse reactions that the vaccine-injured were dealing with . . . but only for one brand of Covid vaccine—AstraZeneca's.

Dr. Nath's paper on the vaccine-injured study that "wasn't happening" at the NIH was published in May 2022, but remained at the preprint stage, without peer review, so wasn't available for use in any official capacity. Nonetheless, the information was there, and again, reinforced the theory that these reactions were related to some kind of immune response, with early interventions being key.

But these scientific papers weren't being publicized. A later Freedom of Information request would reveal that the NIH publicity office was aware of Dr. Nath's study, and they were discussing whether to publicize the study or not with NIH head Francis Collins. They never did publicize it. Physicians weren't being made aware of this new, complicated syndrome that wasn't quite like any other condition they'd seen before. It was like the scientific and medical world had *forgotten* that science and medicine evolves all the time. So neurologists around the country would order the standard MRIs for patients tearfully presenting with new and distressing neurological symptoms, the MRIs would come back normal, investigations were stopped, and the patient was diagnosed with anxiety and sent home with antidepressants.

Despite all of the React19 team's efforts to work with healthcare providers, scientists, the vaccine manufacturers, and government agencies; despite their efforts to share the few papers they could find on adverse reactions past and present with anyone who could actually *do* something with them; it felt like they were being blocked at every turn. All this talk about following the science before the vaccines were rolled out. Where was the science now?

"Fuck them," said the usually calm and collected Joel, "We will find the science ourselves."

Bri, Danice, and Edoardo formed a scientific branch of React19, with the intention of collating the research they were coming across. Research that had been published in mainstream medical journals but was difficult to find. Research containing clues that could be helpful for patients and their doctors; clues about Covid itself, the Covid vaccines, adverse reactions to other more well-established vaccines, and other chronic illnesses. The research led the team to scientists all over the globe, some of whom were very sympathetic to the vaccine-injured in private, even if they felt they couldn't be in public.

React19 published summaries and reviews of all the research they could find, rewriting complex and often monotonous and impossible to interpret papers into language a non-medically trained injured person could understand. The website ended up listing thousands of peer-reviewed studies specific to Covid vaccine adverse events in its "Scientific Publications Directory," and made everything freely available in the hope that patients could discuss possible treatment pathways with their healthcare providers.

And they set up their own patient surveys, sending the results of the first one directly to the NIH. The survey detailed the persistent symptoms of over 500 vaccine-injured people, mostly women aged thirty-five to forty-five. The NIH reviewed the results and requested more information, which the team sent. Then . . . silence.

React19 organized their own research, partnering with Dr. Linda Wastila, who had been an ally ever since the roundtable event in Washington; the study had over 1,300 vaccine-injured participants.

They published a detailed article entitled, "The Spike Protein Problem," featuring an analysis of over 1,300 articles about this spike protein that seemed to play a big part in their ongoing symptoms. The article specifically linked to papers on autopsies, lung disease, myocarditis, severe skin rashes, hepatitis, and multisystem inflammatory syndrome (MIS), the latter being a condition that the injured had begged the FDA to consider as being relevant to them. Some of the UK injured who had been recipients of the AstraZeneca vaccine had been issued a form from the manufacturer about MIS-V (with the "V" standing for "vaccine"), but despite returning the completed forms, had never heard anything from them again.

The React19 team shared all of these studies and details with scientists, medical researchers, and doctors all over the world, in order to spur independent research. Many of these scientists and doctors then used these concepts to inform their work and educate their respective communities on the latest findings. The React19 effort contributed to extensive scientific research around the world, without having the organization's name anywhere on it.

The React19 team shared everything they were doing on their social media pages and in as many support groups as they could and encouraged the global team to do the same. Likewise, the representatives of the injured groups around the world sent React19 everything that was happening in their own countries. It was an extraordinary group effort, all the more remarkable because everyone involved were themselves dealing with debilitating and ongoing health issues.

They were also dealing with the ongoing censorship, so instead of thousands of people seeing vital and life-changing information that could have helped them or their physicians, only a few hundred would. Facebook continued to add warnings to the group's posts, despite the posts containing links to credible, peer-reviewed scientific literature. Kyle was directly contacted by Facebook's fact-checkers—themselves journalist majors as opposed to science majors—and told them that he could answer all of their questions, but he knew they would go ahead and do whatever they wanted anyway. The injured all over the world would find that any links they posted to scientific papers would be shadowbanned, further limiting their reach.

But within that minimal reach, there would be some lives saved. There would be some people for whom simply the knowledge that *some* scientists cared would be enough to give them the strength to face another day. Heidi was never far from Bri's mind.

The React19 website included a section listing any studies, anywhere in the world, that were open to vaccine-injured participants, including independent studies into vaccine efficacy. But the studies were few and far between. Scientific studies cost thousands if not millions of dollars to conduct—perhaps React19 could raise funds themselves in order to provide grants for researchers proposing studies that would improve understanding of Covid vaccine injuries? Surely some organizations or individuals would be willing to fund research into the vaccine-injured? Bri started looking into worldwide funders of scientific research . . . the NIH was top of the list.

At some point in history, had "following the science" become synonymous with "following the money?"

CHAPTER 13

FINDING THE MONEY

With no sign of any reimbursement of the medical expenses and other costs—despite the promises outlined in the consent form—the Dressens had taken the difficult decision to refinance their home. Within just four months, they had gone from being financially comfortable to emptying their bank account, exhausting all their savings, borrowing money from family, and preparing to default on their upcoming mortgage payment. With mounting medical expenses and other costs, Bri now being unable to earn, and no end in sight to the ongoing practical and financial nightmare they'd found themselves in, they felt they had no choice but to release some equity in their home. It was all they had left.

Bri was bedbound when the bank representative came to the house with all the papers. She heard the doorbell ring as she lay in bed upstairs, her stomach sinking to add to the nausea she was already struggling with at that point; and her eyes stinging with tears. *Here's the last of our financial security going, all because of me.*

Brian lifted her out of the bed and carried her downstairs before laying her on the couch, his own eyes stinging as he noted how easy it was to pick her up. There was hardly anything to her; anybody could see that she was very, very sick. Brian's brief explanation to the rep—"My wife is ill"—was unnecessary yet he felt he had to say something. The rep stood observing the sad scenario he had unexpectedly entered into.

He showed the Dressens the refinance agreement, slowly turning page after page and showing them where to sign. The only words he said were "Sign here please," with his voice growing quieter and the atmosphere growing heavier every time he spoke. He didn't usually coordinate the signing of such documents from the floor, sitting next to the signee lying horizontal on a couch. Finally, they were done.

"Are you going to be OK?" The rep couldn't ignore the heaviness in his own heart any longer.

"We don't know," said Brian.

Bri's insurance company eventually agreed to pay some of her medical expenses, but the Dressens were having to find thousands of dollars each month themselves. Brian's ability to increase his own earning potential was limited, given now he had sole responsibility to take care of the family. Being passed over on the promotion he had been working toward for fifteen years had been a shock in some ways, but not in others. His employer expected employees to keep a low profile, and they couldn't have been very happy about the attention Brian's wife was generating.

Some other people were *very* happy about the attention that Bri was generating. As her public platform as an advocate grew, so did the people asking her to endorse their products. Some of them were very influential people with large platforms themselves, and seemingly sympathetic to her plight. She'd agree to have a conversation with them but found their superficiality and aggressive approach distasteful—she'd already tried their products and they didn't work. She told them so, but they didn't care. They said they'd support her financially if she'd just promote their products.

She turned down every single offer, despite her desperate financial situation. She was willing to talk publicly about something if it actually *worked* for her, and did talk about them on her social media in case they could be helpful to others, but she refused to take money for talking about them. She certainly wouldn't endorse a product that she knew didn't work. That whole aspect of being public about her injury and the offers it brought into her life disgusted her.

The only people she wanted an offer from was AstraZeneca.

Following the $590.20 AstraZeneca had reimbursed the Dressens in July 2021, the pharmaceutical company had gone quiet for five months, then offered them $1,243.30. The offer came with the condition that Bri agreed to release them from any further financial responsibility.

After months and months of sending medical bills to them, the offer had shocked Bri. She was lying in a hospital bed at the time, receiving an IVIG treatment that cost $3,500 per session. The consent form she had signed had stated that AstraZeneca would cover all costs associated with any reactions—physical and emotional—that clinical trial participants would have to the vaccine. *They told me they had an insurance policy to cover all this.*

For over a year, Bri had repeatedly missed out on testing and treatments that could have helped improve her condition—Dr. Nath at the NIH had made it clear to her that early interventions were key in the management of adverse reactions. *If AstraZeneca had fulfilled the promises they'd made to me in the consent form, could I possibly have had my life back by now?*

Bri refused to absolve the company of its responsibilities and persisted in sending and resending copies of medical expenses directly to AstraZeneca throughout 2022. But nothing happened.

It wasn't like the pharmaceutical company didn't understand the value of money. Sarah Gilbert, the Oxford Project Leader for the AstraZeneca vaccine, had dedicated an entire chapter on money in the book she coauthored in 2021. Professor Gilbert had secured millions of pounds and a commercial partner in order to develop the AstraZeneca vaccine, saying that she "thought of little else, day and night" during the early months of 2020. She wrote about persuading Oxford University to underwrite the risks as her team plowed ahead with developing the vaccine even without funding secured. She'd lobbied the British government for funds and said that she'd ultimately got £31 million from them. *With all that money, they couldn't set aside any of it for those that were harmed in the clinical trials?*

And then another thought occurred to Bri—a more disturbing one—as she realized the inconvenience of her existence; the inconvenience of the existence of *anyone* who had been severely injured in the clinical trials. *If they'd spent all this money they didn't yet have on developing a vaccine that had turned out to cause injuries, how could they admit that perhaps . . . just perhaps . . . it wasn't as safe as they needed it to be?*

AstraZeneca reported their Covid vaccine revenue for 2021 to be well over $2 billion, and the entire company's revenue for that same year was over $25 billion, up by 32 percent from the previous year. The company provided the vaccine on a non-profit basis until the end of 2021, when the CEO, Pascal Soriot, announced that he no longer considered Covid to be a pandemic, so the product was thereafter to be sold on a profit-making basis, except to poorer countries. The AstraZeneca vaccine was never authorized or approved in the United States and was later withdrawn (by the company's own request) in Europe, by which time it had generated a total revenue of almost $6 billion.

Other companies sold their Covid vaccines on a for-profit basis from the beginning, generating billions in profit for private entities and their stakeholders. Products were developed and manufactured for private companies thanks to significant government funding, generated by the taxpaying public, of which Bri had once been proud to be. *I have paid for these vaccines to be made, now I'm paying for the damage they've caused, and who is stepping in to pay for me?*

And who were the stakeholders in these private companies? Bri was shocked to read that government officials—people directly involved in the decision-making behind Covid restrictions and vaccine development and authorization—also had financial interests in the companies producing the vaccines. Whether or not those royalties were then donated to charities felt irrelevant to her—it was all still income. *Shouldn't we have been made aware of that? Should such a conflict of interest even be allowed?*

It wasn't just the vaccines themselves that had received considerable funding. The rollout had cost money too, leaving even more questions . . . How much did the clinics get paid? How much did the doctors who administered the shots get paid? What about the people who pushed the injections into people's arms?

Then there was the advertising campaign. How much did social media companies get paid? What about the "fact-checkers" that were censoring all the vaccine-injured and shutting down the support groups? How much did the NFL get paid to host Peter Marks with all those famous players on that podcast where Sheryl had had her questions about injuries removed? How much did the person who removed Sheryl's questions get paid? What about the celebrities? Arnold Schwarzenegger, Martha Stewart, Samuel L. Jackson, Jane Fonda? Who paid Stephen Colbert to do that dance promoting the vaccine? Who paid for his performers, their costumes, and the choreographer? *Who paid for all this? How much did it all cost?*

It had cost Bri her life.

With all that money flying around the Covid vaccine—let alone what was invested in the management of the pandemic overall—nothing seemed to have been allocated to those who were harmed by it.

Civil rights and vaccine attorney Aaron Siri had proposed what seemed to be a reasonable solution, in a letter he sent to HHS, the CDC, and the FDA. The letter was dated October 27, 2021, and included testimonies from the injured, many of whom had already been directly in contact with Dr. Woodcock and Dr. Marks. By that time, both government employees had been made fully aware of the impact of the injuries on multiple people's lives.

Aaron suggested that 1 percent of all Covid funding be made available for research, prevention, and treatment of Covid vaccine injuries. If the figures later reported in the US Government Accountability Office website were accurate, then Aaron's recommendation would mean that of the $4.6 trillion invested into helping the nation respond to and recover from Covid, $46 billion would be allocated to vaccine injuries.

Instead, the vaccine-injured were being told to apply for compensation through the CICP, a notoriously difficult process that had to be applied for within one year of injury—many of the injured weren't even able to face the process in their first year of illness.

As of January 1, 2024—almost three years after the majority of injuries occurred—12,854 Covid vaccine claims had been filed in total. Just eleven claimants had actually received any money; the average payout was around $3,700. The United States House of Representatives Select Subcommittee on the Coronavirus Pandemic conducted an investigation into the country's vaccine safety systems; soon after there followed a surprise outlier payment of $370,376 to a single claimant. Of the 3,130 claims that had been completed by August 1, 2024, over 98 percent were denied.

Claims were awarded for myocarditis and myopericarditis (inflammation of the heart), syncope (loss of consciousness), and anaphylaxis (allergic reaction), all of which were recognized adverse reactions to the Covid vaccine. There existed a backlog of over ten thousand claims that were either pending review or being reviewed, and the program had to increase its staff from four to over thirty due to the influx of claims after the Covid vaccine rollout. It had cost over $26 million to manage the program—$26 million to pay out just over $420,000, with the vast majority of that being to just one claimant.

The system was widely reported in the US media to be not fit for purpose. And the same was happening with similar systems all over the world.

> "So who can we sue?" Bri, Sheryl, Danice, and Kristi were on their first phone call together with a lawyer.
>
> "Nobody."
>
> "What about the drug company?"
>
> "No."
>
> "The hospital who gave me the shot?"
>
> "No."
>
> "What about the government?"
>
> "No. The PREP Act protects all of these from any legal repercussions."
>
> "Every one of them is protected?"
>
> "Yes. All of them. You cannot sue *anyone*."

They were all silent, trying to process what the lawyer was telling them. Sheryl wiped a tear from her eye, then another, then another. There was no way out of this. They had lost their health, they had lost their income, they were on the brink of losing their homes, and there was absolutely nothing they could do. They hadn't known that when they accepted the Covid vaccine, they accepted *all* of the risk associated with it too. They had all gotten vaccinated oblivious to the fact that their rights have been removed. Unlike any other product available to the public, there was nothing in place to protect them—the consumer—if something went wrong. Meanwhile, the vaccine manufacturers were making a fortune. *How has this happened?*

It happened because, for decades, a fundamental change had been taking place in American healthcare that meant it was no longer a patient-first endeavor; it was business-first. It had gone from not-for-profit to major profit.

Hospital administrators who were once doctors hired because of their decades of hands-on experience were replaced by smart suits and business qualifications rather than their knowledge and bedside manner. Doctors who once ran clinics based on the needs of the patients they served had become mere employees, with strict instructions on what they were and were not allowed to do and say; there was no room for exploratory investigation for any patient that did not fit into a flowchart of predetermined symptoms. Chronic disease affected more Americans than any other illness, and often came with multiple symptoms and syndromes that meant boxes could not be easily ticked. Investigations and lengthy office visits would not make anybody money. So the sooner a chronic illness could be diagnosed as something *treatable*, then the sooner the visit could be billed to insurance, and the sooner money could be made. As chronic illnesses often impact a patient's mental well-being, then those boxes could be ticked, and antidepressants or anti-anxiety medication could be prescribed. Next patient, please.

Large corporate hospital incentive programs meant that doctors who had begun their careers seeing ten patients a day would be ending their careers with fifty a day, and threats of pay cuts if they couldn't take on more. There was no time to be reporting new and concerning findings to peers and superiors, and no structure set in place for those peers and superiors to gather together in collaboration to decide on the best way forward for their own patients or other clinics. There was a reason that chronic illness sufferers around the country felt like an inconvenience to their doctors; it's because they were.

Small family clinics were being acquired by major hospital systems. Ranking systems applied to the employee doctors as well as the healthcare facilities themselves. Corporate hospitals met an obese patient—who posed a higher likelihood of dying during surgery—by telling the patient to come back after losing some weight. Fatalities were not good for the hospital's ranking system. If the patient dies at home trying to lose weight, the hospital's corporate ranking system would not be impacted.

The healthcare model was no longer bottom-up; the doctor reporting up to the health agencies. It was now an efficient business-first model, from the health agencies down to the doctor. Healthcare had become a huge money-making industry, and doctors had become employees working within that industry. And just like employees in any other profit-generating business, doctors were constantly under pressure or incentivized to meet deadlines and goals rather than prioritize the needs of the patient. Patients were being denied the humanity that had been the original motivation for so many doctors and nurses to enter the medical profession in the first place. Doctors and nurses were being denied the ability to do their jobs properly—jobs they had probably dreamed of since they were children.

Medical expenses skyrocketed under this business-first model. New drug prices increased from an average of $2,115 per year in 2008 to over $180,000 per year in 2021. A 2016 article on prescription drug prices concluded that the most important factor that allowed pharmaceutical companies to set high prices was the FDA's protection of the companies' exclusivity through monopolies and patents.

But what did it all matter, if the insurance companies were footing the bill?

Except they weren't. The cost of healthcare was just being passed on to the patient as insurance companies charged more and more for premiums but paid less out, and the patients' out-of-pocket expenses increased. Patients were just one health condition away from losing their homes, because a health *service* had become a health *industry*.

Part of that industry involved a multi-billion-dollar lobby, ensuring that successive governments had slowly eased restrictions and regulations on pharmaceutical companies and large corporate healthcare systems. The American Medical Association published a study showing that healthcare lobbying had increased by 70 percent in the twenty years preceding Covid; pharmaceutical and health product manufacturers had spent over $300 million on it in 2020 alone.

This is the system into which Covid landed. A system where hospitals were incentivized by certain treatment protocols over others, doctors were discouraged from investigations of complex cases (which all virus- and vaccine-related illnesses were), and politicians were pressured by the pharmaceutical industry to do what was best for business. Overshadowed by the fear and panic that the media created around Covid meaning that most normal people were so concerned with their own health and that of their loved ones that they weren't even aware of what was going on with the health system. Until it was spelled out for them.

Democrat or Republican—it didn't matter who had been in charge. It didn't matter how anybody had voted. Everybody could unite in all facing the same problems with the healthcare system. From that moment when Bri had sat in the clinical trial waiting room, checking her phone for the election results that would send Trump out and bring Biden in; nothing had changed. The lobbying had only become more powerful.

Kyle Warner had seen firsthand how pharmaceutical lobbying worked in Washington. Try as he did to return to the world of professional mountain biking, Kyle's heart wouldn't let him—either physically or emotionally. That first trip to Washington where he heard the stories of all the other injured who spoke at Senator Johnson's roundtable event had changed him. It had changed all of them. Kyle couldn't walk away from his new friends, even if his body would let him. So he'd taken on the role of React19 lobbyist. He accompanied Bri, Joel, and the other injured advocates on multiple trips to Washington, DC, and talked to numerous senators. He noticed that there would often be two representatives from the pharmaceutical industry either leaving the office just before his meetings, or coming in directly after him.

The leftover funds from the first roundtable event in Washington had been the start of React19's serious fundraising efforts in November 2021. Among other projects, they wanted to create a network of providers. A small but growing number of healthcare providers had begun to treat vaccine injury, but access to these physicians was limited by geography, insurance, and capacity. With the right connections and money, React19 could help the injured overcome these barriers.

The influential people with money and a platform of their own who *seemed* to be sympathetic providers were making money from the desperate vaccine-injured, some of whom were willing to pay anything for a product that came with promises of a cure. But when faced with having direct conversations with the injured at the rally, the advocates had been ignored by those who professed to be helping them.

The events in the spring of 2022 were completely different. By then, everybody attending wanted to know more about React19. Bri, Joel, Kyle, and the other board members were no longer invisible. Other attendees actively sought *them* out. Robert F. Kennedy Jr. was in regular contact with Bri by then, making clear his genuine support for the organization, which influenced others at events. And the React19 advocates had grown in confidence. The January rally had shown them just how much support was out there from *ordinary people*; people who donated what little they had because they were so moved by the stories that had been bravely shared on the steps of the Supreme Court on that rainy day. The love that had been shown in even the smallest of donations had given the advocates encouragement that far extended the monetary value. Bri and the others were determined to connect with anybody that could help them practically, medically, or financially.

"Hey, we are React19. Are you a doctor? You should register with us." The venues became abuzz with the same words repeated over and over again, as the advocates collected business cards and contact information from medical professionals, healthcare specialists, and potential donors. They made enough contacts through the spring events that they were able to launch their treatment provider network, while all continuing to volunteer their own time.

Despite their own loss of income; neither Bri, Joel, or Kyle—or anybody else on the React19 board—personally received any money for their efforts. There wasn't anybody at the head of this particular charity making money from its activities. All money raised through public events and the website was donated—and still is—to React19's very own "CareFund."

Many of the injured throughout the world had launched their own personal healthcare fundraisers; often self-censoring so that the word "vaccine" wasn't even mentioned, so conscious was the community of the risk of having the ability to fundraise pulled.

The healthcare situation in the United States meant that Americans weren't unfamiliar with seeing personal healthcare fundraisers; the CEO of one of the largest online fundraising platforms, GoFundMe, had been quoted in 2019 as saying that a third of the platform's fundraisers were for medical expenses, something about which even he had been surprised. A handful of the injured who had relatively high profiles were able to bring in modest amounts of money toward their healthcare expenses, but without notoriety or a bigger platform, the majority of individual injured's healthcare fundraisers would only bring in a few hundred dollars.

But React19 made the decision to stay away from individual fundraisers—not wanting to be accused of nepotism or to participate in any activities that benefited just one or two individuals who had taken the step to be very public about their injury. React19 wanted their fund to be open to anyone in need, irrespective of their public status.

So the React19 CareFund was established for anyone resident in the United States in financial need. It was intended to cover past medical expenses that insurance companies had refused to pay for, or treatment that a vaccine-injured person needed but couldn't afford or get coverage for. The fund wasn't set up to go toward vacations, or even to help someone with their rent. It was purely to help with out-of-pocket medical expenses. Donors to React19 could specify that their donation went into the CareFund as opposed to other React19 activities like research funding or advocacy campaigns, and board members would seek separate funding for those efforts.

The application process involved the consideration of the applicant's financial situation such as income and employment status, household assets and liabilities, outstanding medical bills, and insurance status; as well as their medical situation. Most people applying had no savings and no income. Even a part-time job or disability income would disqualify someone from applying. The fund was for people in very, very desperate situations.

The applicant's diagnosis, treatment to date, recommended future treatment and associated cost were also considered. The CareFund didn't restrict the types of treatment to pharmaceutical only; all kinds of treatments were considered valid—energy healing, acupuncture, trauma work—but conditions that had clearly existed beforehand did not qualify.

The application then had any personal identifying information removed before it was submitted to a committee for review.

Hundreds of people applied, and all but a handful were successful, with the average payout being about $6,000. Over $850,000 was awarded in the first two years. There was always a waiting list and there was never enough money coming into the fund so it would periodically be put on hold. It was becoming increasingly difficult to secure donations as the world moved on from the chaos that was the Covid era, and thus the vaccine-injured shifted to the back of people's minds.

Even the most sympathetic and understanding allies found it difficult to believe just what little government support the vaccine-injured were receiving. And just how much the injured were *still* struggling, years after being injected. It was often in the years after injury—rather than the months—when the financial pressure of trying to manage a condition that simply wasn't going away, would really take hold. Savings would all be gone, the generosity of family exhausted, houses remortgaged or sold altogether, along with anything else of any value.

The React19 CareFund was, for some, their very last hope.

Maddie's parents had put everything they had into trying to get their daughter better. Maddie had unwittingly become the poster girl for vaccine injuries, and Steph had been inundated with promises from "experts," claiming that they could get Maddie walking again, or back to being able to swallow food. Steph had put thousands and thousands of dollars on her credit card traveling all around the country to try various treatments. Nothing had worked.

IVIG—the treatment that was helping Bri—remained elusive without a medical professional willing to prescribe it for her, but Steph had been abandoned by the Ohio hospital system tasked with caring for children. This made her very vulnerable to being surrounded by people who had told her that IVIG would poison her daughter. One doctor from the freedom movement had just a single conversation with Steph about Maddie's care, and then proclaimed to others that they were now treating Maddie, something which would do wonders for their image. Like many of the injured and their loved ones, Steph had no idea who to believe or trust anymore. Nowhere to turn, no money in the bank, and a daughter whose condition was not getting better—she spent night after night hiding in her closet so that nobody would know she was crying, completely overwhelmed and broken, with Bri on the other end of the line.

Like Bri, Maddie had been completely abandoned by the pharmaceutical company in whose trial she had participated. The Cincinnati Children's Hospital system that was supposed to care for Maddie had reported her injury to the FDA as a mere stomachache and told Peter Marks they didn't "feel" her injury was related to the vaccine. This same hospital system was being paid by Pfizer to continue to run the children's clinical trial and would later announce the opening of a new, sizeable, state-of-the-art testing facility for more children, paid by Pfizer.

Bri told her to apply for the CareFund, but Steph couldn't bring herself to. She was a gentle, humble woman, and the refusal wasn't because of pride; it was because she believed that there were others who needed it more. She didn't want to put her daughter's suffering above anyone else's. Bri's father offered to help. Steph couldn't bear the thought of taking his money.

Similar kindness was happening all over the world as people who were connected to the injured community stepped up for each other financially as well as emotionally. If somebody came into a little windfall, they shared it with their new friends. Healing gifts or treatment vouchers would arrive in the mail, addressed to someone who was right in the middle of their own darkness, from someone who had recently emerged from their own.

That phrase used by English-speaking governments the world over— "We're all in this together"—was never more true than within the vaccine-injured community.

Finding specialists to treat vaccine injury was challenging enough, but finding specialists to treat vaccine-injured *pediatric* patients was next to impossible. Steph and Bri searched high and low for providers to help Maddie. Joel finally connected Steph with a new React19 provider, Dr. Molly Rutherford. From her unassuming family medical practice, she was treating some of the most complex conditions in the post-Covid world, proving that a board-certified specialist wasn't always necessary. Dr. Rutherford learned just how much doctors could gatekeep someone's ability to access medical treatments. She prescribed IVIG for Maddie and wanted no recognition in return. With savings at zero, and funding Maddie's care with credit cards, Steph finally agreed to apply to the CareFund, and Maddie finally received a major treatment that turned around the direction of her health.

More than three years after participating in Pfizer's clinical trial, Maddie finally started to experience the return of her sensory nerves. The painful process brought sensation back to her waist and legs. Steph was simultaneously encouraged by her now fifteen-year-old daughter's slow improvement, yet also devastated at the thought of the damage that had been done through delayed diagnosis and treatment.

Maddie wasn't the only child who had a harrowing experience with the Covid vaccine. Ryleigh was vaccinated in January 2022 and diagnosed with Covid six days later; left with blackouts, numb limbs, memory loss, and incontinence. This former bubbly and lively eight-year-old became wheelchair-bound. All tests were clear, but Ryleigh was in so much pain that she couldn't bear to have anybody touch her. Within weeks of becoming ill, little Ryleigh was eventually committed to a psychiatric ward and prescribed antianxiety medications.

Ryleigh's mom found Steph online, which led her to the tools that React19 had put together. Steph told her of early intervention measures as indicated by the NIH. So Ryleigh's mom decided to take her daughter out of the psych ward, in which she was only declining. Upon her exit, the hospital reported Ryleigh's mom to Child Protective Services, but she had successfully applied to the CareFund and was able to fly Ryleigh out of state to a physician who was part of the React19 underground provider network. The doctor prescribed IVIG.

Within weeks Ryleigh's health had returned to its original state, and she walked herself back into school that fall.

It was another example of Dr. Nath's hypothesis that early intervention with immunotherapy was key.

And despite the NIH team knowing all about the vaccine injuries—and having theories regarding how to treat them—Ryleigh's mom's testimony given directly to the FDA was dismissed by Peter Marks in Washington later that year. Bri explained to Dr. Marks that if it were up to the government, Ryleigh would still be in diapers and a wheelchair, stricken with an illness that is "misinformation." Instead, her mom connected with others who knew the secrets of the NIH, leading Ryleigh to walk into school as if nothing had happened to her.

Dr. Marks had said, "There are no secrets here."

It wasn't just children who were having their lives changed by the CareFund and React19's provider network. The organization was changing adults' lives too. "Because of you, I was able to get to a doctor who actually listened." "Thanks to React19, I can seek medical care beyond the compassionate assisted suicide my provider offered me." "For the first time ever, I felt like a human."

The React19 website was full of testimonials from people expressing their relief at being recipients of the CareFund, with many saying that the news had brought them to tears. They could finally access tests outside of the US as well as domestically, try that treatment they'd been desperate for, or resume a treatment that they already knew worked for them but they'd run out of funds for it. The testimonials were a deeply moving read.

Nicki Holland had gone from being an active physical therapist to spending almost a hundred days in a hospital. Her condition was so volatile she was continually life-flighted to another ICU; one life-threatening event after another. She was confined to a wheelchair with both breathing and feeding tubes. She was doing everything she could with minor medication and natural interventions but continued to be dramatically unstable. This young single mom of three desperately needed something to change. She applied to the CareFund to have treatment covered, and that treatment started her on a very slow path to healing, allowing her to breathe independently, and her health to stabilize, putting an end to the life-flights.

Louie Traub was living in his car when he applied for the CareFund. A talented photojournalist, he had sold all of his equipment to pay for treatment and was living in parking lots after becoming unable to pay his rent. While the fund wasn't awarded to cover rent or general living expenses, it could at least cover his medication, and he eventually changed states to live with his parents.

Not all the injured had family support. One severely injured woman in Ohio was bedridden with progressing neuropathy, similar to Bri during the early months, but unlike Bri, she didn't have support at home and her condition was putting significant strain on her marriage. This young mother of two repeatedly went to the Cleveland Clinic hoping for medical help, but instead they offered her assisted suicide. She confided in Bri that she had decided to go ahead with it; she couldn't bear being such a burden to her family anymore. She applied for, and received, the CareFund. She never got the support she needed at home, and her husband put her in a rest home, but she didn't follow up on the clinic's offer to help her end her life.

Unlike another injured woman in Canada, whose husband did leave, along with the adult children, who he told would end up spending their lives caring for their disabled mother. She chose Canada's assisted suicide program.

Countless previously active long-term partners—life-creating and life-loving—in becoming vaccine-injured had through no fault of their own also become at risk of what was legally considered "criminal abandonment"; the act of a partner withdrawing all practical, emotional, and financial support for a sick spouse. Most injured—and their partners—had no idea this was considered a criminal offence, and that a court could require the abandoner to maintain such provisions to the person they were walking out on.

Bri had stopped counting the suicides when they reached twenty-seven. But she did note that all except two of them didn't have support at home. Facing the medical gaslighting, the media blackout, the lack of government recognition, the debilitating symptoms, the financial pressure, and the loss of identity without someone at home who even cared, was understandably too much for anyone. She didn't blame anybody for ending it all. Without Brian's support, she would probably have done the same.

The React19 CareFund sometimes quite literally saved people from pursuing what they felt was their only other option.

It wasn't like the CareFund could solve *everything*. There wasn't enough money, there were too many people in need, and the situation was too complex. But the CareFund could give recipients the opportunity of trying a new treatment that might give them a bit of relief; it could give them access to a sympathetic healthcare provider whose kindness *alone* might restore just a bit of faith in humanity again; it might lead to feeling like you'd found somebody who cared. It gave people a reason to keep going.

The CareFund was about so much more than just the money.

CHAPTER 14

FINDING TREATMENTS

The vaccine-injured who had taken the step of going public with their condition and the growing community that they had unwittingly joined, were frequently accused of being "grifters." Social media was awash with individuals who confidently criticized and ridiculed anything that was said by vaccine-injury advocates . . . yet weren't confident enough to use their real names while doing so. Bri, Joel, Kyle, and the other React19 board members, as well as other prominent injured throughout the world, were subject to online abuse and accusations of being just out to make money from the government or the vaccine manufacturers.

The accusations showed an incredible ignorance into how the vaccine industry worked and a lack of knowledge of just how difficult it was to get any kind of money to replace or compensate that which had been lost through earnings, let alone the new expenses the injured found themselves facing as they attempted to manage their symptoms. The accusations also gave the injured yet another layer of trauma to deal with on top of their injuries, and they would all sometimes ask themselves just how much it was costing their own health by being public in order to help others with theirs.

If you gave any of the injured the choice between money or getting their health back, all of them would choose their health. To the vaccine-injured, money was merely a tool to enable access to treatments; treatments that could be found by the most persistent and motivated within the community themselves.

Some of the injured realized and accepted very early on that their doctors were unable to help them. Some had become so disillusioned and mistrusting of the pharmaceutical industry and those who participated in it, that they didn't even *want* their doctors' help. It felt like doctors around the world had become the gatekeepers to the health of the vaccine-injured—even those medical professionals that *were* sympathetic were beholden to healthcare systems that restricted access to diagnostics and treatments unless specific criteria were met. The vaccine-injured often didn't meet those criteria. If testing *could* be accessed then the tests often came back all clear and instead of saying "we don't have the tests to help us understand your condition yet," doctors would say, "there's nothing wrong with you."

But the vaccine-injured knew that something was very, very wrong with them.

So some decided to take responsibility for their own treatments, and spent what little energy and whatever money they had navigating through information and genuine allies that *were* available, in order to create their very own individual treatment pathways and protocols. At least in America, money meant that the gatekeepers could often be bypassed, React19 could continue to develop its underground network of healthcare providers, and treatments could be accessed.

For some of the injured, taking control of their own recovery was incredibly empowering.

While some treatments were sought out to address very specific symptoms or organs, such as gabapentin for nerve pain, colchicine for pericarditis, or EMDR for PTSD, most of the injured were dealing with systemic problems that affected them literally from head to toe. Treatments were needed to support the entire circulatory system, lymphatic system, and digestive system. In some ways, a simple approach was required that involved looking at what went *into* the body, what happened to anything after it went in, and what was eliminated afterward.

Diet was one of the first areas that many of the injured, including Bri, changed; sometimes by accident and sometimes out of necessity. With a sudden onslaught of intolerant and sometimes allergic reactions to foods that the injured had previously enjoyed with abandon, for many of them their diets had become very restrictive. Bri had accidentally discovered that eating unprocessed foods had reduced her symptoms, and she ended up completely eliminating sugar, dairy, wheat, and corn as well as anything processed, always taking her own fresh fruit and vegetables when traveling. It became a standing joke among the injured that Bri would get stopped at airport security because she had potatoes in her luggage.

Injured throughout the world were experimenting with low histamine, anti-inflammatory, anti-candida, SIBO (small intestinal bacteria overgrowth), anti-mold, and anticoagulant diets, as well as with Keto or carnivore diets. The most nutrition-focused created their own unique diets as they became more in tune with how their bodies responded to foods, and in some cases, specifically used certain food items as medicine. People who had never thought twice about their food became familiar with the names of every ingredient listed on the back of any food product, switched to eating only organic to minimize further toxic exposure, learned how to make their own bone broth to heal their gut linings, and could tell within minutes of consumption whether they had accidentally eaten something "unsafe." Nutrition science had been studied for over a hundred years— there were plenty of research papers and nutrition experts from whom the injured could get inspiration and information if they were making their diet a key component of managing their own health.

It wasn't just what to eat that was becoming key for many; some were experimenting with not eating at all. Intermittent fasting and even days-long fasting was tried by some, with the intention of bringing about "autophagy"—when the body consumes its own damaged cells and prompts the creation of newer, healthier cells.

Pooping and peeing habits were monitored, both of which had been affected in the weeks if not months after the vaccine, and for some continued to be problematic in the years post-jab. The contents of both poop and pee were examined by the injured themselves if not by gut specialists, as those who had decided to take control of their own recovery became more knowledgeable about indicators that the body was functioning as it should.

People learned about the lymphatic system—responsible for the removal of waste not eliminated by the digestive system. Lymph—unlike blood—doesn't have a "pump" to ensure its flow around the body, and relies upon movement or massage to keep it going. There were theories that the vaccine had exerted great demands on the lymphatic system, and with many of the injured no longer being able to exercise, it was thought that the system was becoming clogged up. Bri had this confirmed during a massage with a sports therapist who had been familiar with her body prior to vaccination. He had said that her spleen (part of the lymphatic system) and the lymph nodes around her armpits and breasts were now rock hard, and he worked on unblocking them. Bri felt her symptoms temporarily subside right after the treatment. Body brushing, lymphatic drainage self-massage, and any activity that encouraged the body to sweat (such as saunas) became regular activities for those who were keen to keep this valuable waste-processing system at optimal functioning.

And then there was the circulatory system. Research in Germany and South Africa had indicated the presence of "microclots"—tiny clots that could not be detected by normal clotting tests—in the blood of the vaccine-injured and Long Covid patients, with theories that the spike protein was the cause. One of the NIH researchers had discussed the possibility that Bri might have microclots when she'd visited them in 2021.

Within weeks of the AstraZeneca rollout there had been reports of sometimes fatal clotting disorders, prompting the halting of the availability of that particular brand of vaccine in some countries. Some already injured were experiencing very similar symptoms to those who had died, as well as seeing their blood behave differently than it had before the shot. The support groups were full of people talking about difficulties phlebotomists were having in getting any blood out of their previously cooperative veins, and that when they could get blood out, the blood was so thick and dark that it was clogging up the tubes. People were finding that when they did an at-home fingerpick test, they couldn't even get a drop out on to the test . . . the blood coagulated right there on the tip of their finger.

Some of the injured talked about initially being fine post-vaccine, then suddenly ending up in the hospital with a blood clot, despite not having any risk factors. That clot would be dealt with, and soon after, there would be another one, then another, with doctors having no idea where these clots were coming from or why, having no success with heparin treatment, and not knowing that there was research connecting these clots to the spike protein. But the vaccine-injured knew.

People tried foods and supplements that were considered to "thin the blood," as well as over-the-counter medicines such as aspirin, and even took pharmaceutical anticoagulants if they had a doctor who was willing to prescribe them.

But the most crucial tool that Bri—and many holistic practitioners—believed the injured had at their disposal, was something very simple . . . sleep. If only the injured could get any.

Constantly exhausted, the early months for most of the injured were plagued by sleep problems. Either they were sleeping too much, or not enough; sleeping during the day and wide awake at night; going to bed tired, only to find their bodies and minds come alive the moment their heads hit the pillow. Severe nightmares, vivid dreams, and insomnia were normal for most of the injured. Nobody was waking refreshed and ready for the day ahead. For some, these sleep problems went on for *years* post-vaccination.

Bri had been obsessed with sleep and food during 2021. *If I can just sleep then I can survive. If I can just eat then I can survive. Sleep. Eat. Sleep. Eat. That's all I need to do.* She imposed a strict but simple house rule—if it was dark outside then all the inside lights were turned off, sending a message to her body that it was time to sleep. All electronic devices were switched off, including the kids' devices. It took six months, but by implementing strict routines and having a variety of medications at night, her sleep pattern returned to the typical seven hours she had had before the shot. Anytime she didn't allow her body time to calm down for a few hours before bed, she was back to square one again with a multitude of nocturnal issues, which then in turn made any other treatment a waste of time and money. Sleep was the number one tool she had at her disposal, and it had to remain a priority. That, and drinking lots and lots of water.

Bri felt that it was only after getting the basics under control—sleeping and eating as well as the elimination of cellular or digestive waste—that any pharmaceutical intervention had any chance.

While some of the injured went strictly natural in their treatment pathways, and some were heavily reliant on pharmaceuticals, Bri incorporated both approaches in her own protocol. IVIG had brought about incredible relief of her symptoms, as had a drug to calm her immune response, two different kinds of antihistamines, and steroids as needed. But the prescribed products weren't for everyone—many of the injured found that they not only reacted to food, but they also reacted to pharmaceuticals, at least in the early stages of illness. Bri herself had had reactions to medications she had tried since becoming injured, and even passed out following her first dose of IVIG. But she'd felt that it was worth it for the relief it had brought her.

She often thought about the price she would end up paying for taking all of the medications she tried, none of which came without side effects, and some of which would actually be dangerous to take long-term. Every time she tried to come off something her symptoms would no longer be manageable, and she struggled with accepting that she might be on pharmaceuticals for life, even if what she took would ultimately shorten that life. She felt that she was buying time from her future, in order to be present *now*; she told herself that her young kids *needed* her now, and that was what mattered.

Either because of the vaccine or because of all the drugs she was now on, she might not see her kids get married, but she might see them graduate high school.

Bri was learning to accept a very different life than she had envisioned for herself, and that included learning to accept treatments that had been ridiculed by the mainstream media in which she had trusted before becoming ill. Hydroxychloroquine was a good example. The anti-inflammatory drug had been derided as a "fishtank chemical" in the news, and the idea had stuck in her mind, so she had refused any suggestions of trying it until almost four years after becoming injured. She tried it during an especially bad flare-up of her original symptoms, was surprised at how effective the treatment was for her, and then annoyed at how she'd resisted it for so long, just because of what the media had said about it; a media that she now knew wasn't being honest about the vaccines. *What else was it not being honest about?*

And there were plenty of other products the vaccine-injured were trying that didn't require the cooperation of a prescribing doctor.

Bri herself experimented with a range of vitamins, minerals, and supplements and added some of them to her own protocol. She took supplements for circulation, nerve pain, and brain function, as well as antioxidants, digestive enzymes, and electrolytes. Sometimes she went back to things that didn't work for her during the acute phase of illness, only to find that they became extremely helpful to her much later down the line. She tried to keep an open mind and to remain proactive with her own recovery.

Kyle Warner was a great example of what a proactive and highly motivated vaccine-injured person could achieve with out-of-the-box thinking. He approached his own healing with the same attitude with which he had always approached his mountain biking training, and focused all his energy on whatever treatment he felt could work for him. He got the basics under control—sleep and nutrition—then tried a few pharmaceuticals but ultimately focused on an antihistamine diet, vagus nerve reconditioning through ice baths, and an extensive series of hyperbaric oxygen therapy (HBOT) sessions. HBOT involves breathing oxygen while being put in a high-pressure chamber and is used in the US to treat decompression sickness and poisoning, among other conditions. Kyle's protocol didn't return him to his previous athletic condition, but as long as he kept his heart rate down, he was able to do some exercise. He was one of the very few people Bri knew who could be relatively active, but it wasn't without constant effort and awareness on his part. It was a far cry from the level of activity he had enjoyed as a professional athlete, but it was something.

It was a constant effort trying to work out which supplement or protocol would be helpful and which was a waste of money. Support groups were full of members hopefully sharing videos of the latest "cure" being touted by people who had become household names in the non-pharmaceutical world, and admins were constantly having to deal with unscrupulous providers wanting access to their vulnerable members; providers who participated in global conferences and podcasts as if they knew all about the vaccine-injured, when in reality they'd just had a brief conversation with vaccine-injury leadership.

The fearmongering that had contributed to many of the injured choosing to get vaccinated in the first place now continued for this marginalized section of society. Just as the world had been told that the vaccine was the only way of combatting illness caused by Covid, now the vaccine-injured were being told that this product or that product was the "only" way of combatting illness caused by the vaccine. The rest of the world had opened up, no longer afraid, but the vaccine-injured remained in a state of fear that was actively encouraged by the very doctors claiming to be helping them.

Those at the helm of the support groups around the world would call out such claims. Bri spoke directly to one doctor, telling him repeatedly that he couldn't cure the injured and he had to stop claiming as much. He'd eventually conceded, "You're right. I can't cure them. There is a lot we just don't know."

Treatment of the vaccine-injured was a highly exploitative market. Proprietary formulas of products that alleviated some symptoms were registered, leading to the products' availability and pricing being controlled by certain individuals. Bri would have preferred that those individuals worked with the injured communities to ensure that every person had access to something that could give them just a little bit of their lives back.

This exploitation was indicative of what could happen when the government wasn't recognizing, researching, or regulating the treatment of the vaccine-injured. It was indicative of what could happen when vaccine injuries were censored. Many of the injured saw someone in a white coat talking about vaccine injuries on social media and, desperate to be cured and having just been dismissed by their own doctor, assumed that this friendly, sympathetic one on a podcast with thousands of views *must* have the answer.

But there was no "one answer." Like many complex health conditions, one size did not fit all.

Bri's neurologist, Suzanne Gazda, told her that what Bri and her new friends were experiencing was the most complex of conditions she had seen in her entire career. Dr. Gazda had dealt with MS, ME/CFS, ALS (amyotrophic lateral sclerosis), and Long Covid, as well as many other neurodegenerative conditions, for thirty years. Treatment required the kind of experimentation that some individuals within the injured community were focusing on; just as other chronic illness communities had been experimenting with for years beforehand, always bearing in mind that every individual was different.

There were enough similarities between the Covid vaccine-injured, and people with ME/CFS or Long Covid, that some of the treatments used for the latter were worth experimenting with. Both the ME/CFS and Long Covid communities—along with other complex chronic conditions such as POTS, MCAS, fibromyalgia, mold-related illness, autoimmune conditions, Lyme disease, mitochondrial dysfunction—had been battling to achieve recognition and research that could lead to possible treatments, had their own established support groups, and in some cases their own charities and even their own dedicated clinics. Much could be learned from these other chronic illness communities.

Some of the support groups were open to discussions around the possibility of past vaccination (of any kind, not just Covid) playing a role in the development of their own conditions; some were highly censored about the subject, and the vaccine-injured again self-censored, this time in order to learn about treatments that could help them.

It was within these other chronic illness support groups that the injured learned about hydration and specifically salt intake, pacing and "spoon theory," eliminating mycotoxins, and limiting exposure to everyday chemicals in the home. They learned about many different components of food and the natural chemicals within the body that were related to certain kinds of foods being consumed—histamine, tyramine, digestive enzymes. They learned about what was added *to* foods in order to preserve them, protect them, enhance the flavor, or make them more desirable and/or addictive.

They learned about tests that could assist with getting *some* kind of diagnosis; even if healthcare providers refused to acknowledge any connection with the vaccine—at least the test results could lead to possible treatment. They learned about research into similar conditions, and papers that had been peer-reviewed and published in credible scientific journals, giving them something they could print out and hand to their own doctors. The chronic disease community inspired React19 to conduct their own focus group, submitting member samples to a specific lab for testing— they all came back with evidence of mitochondrial dysfunction, something for which pharmaceutical and natural treatments were available. The cells' "batteries" weren't working properly; and there were pharmaceutical and natural treatments available for mitochondrial disease.

The Covid vaccine-injured also learned from people who had been injured by past vaccines, and from the providers who treated them. They had all been subject to similar censorship so they were hard to find, but many of them reached out directly to the Covid vaccine-injured community. A small clinic in the northeast of Japan had successfully treated teenage girls who had been injured by the HPV vaccine; another clinic in Denmark had had the same success with a different method.

Around four hundred Long Covid Care Centers were established in the United States, which was helpful to some, but the vaccine-injured—despite the similarities in symptoms—weren't necessarily given access to the centers despite the similarities in symptoms. The NIH had received over a billion dollars from the federal government to study Long Covid, and research papers could again be helpful to some, but the studies were often emphasizing the impact of Long Covid and minimizing or outright dismissing Covid vaccine injuries. These papers with the high price tag quickly became a source of criticism among the Long Covid communities as well. Sometimes it felt like these science experiments existed simply to encourage people to get vaccinated.

So the vaccine-injured became their own "science experiment."

Beyond the basics—sleep and nutrition—and whatever treatments that could be accessed for specific symptoms or diagnosed conditions, the injured explored anything and everything that could bring about some relief. The support groups were full of people asking if anybody had tried this treatment or another, and the admins frequently ran polls to gather more information. The admins of the support groups became incredibly knowledgeable on what was helping the Covid vaccine-injured; but nobody was actually asking *them* what worked and what didn't.

Some were finding relief by using red light therapy, frequency devices, frequency music, sound baths, meditation, grounding, prayer. Some found relief from acupuncture, reflexology, kinesiology, spinal realignment therapy, yoga, deep breathing, cold showers. Many experimented with nicotine—patches, gum, or vaping—which had scientific papers published supporting nicotine as a treatment for Long Covid. The effects of copper on coronaviruses had been studied by scientists long before Covid, with Professor Bill Keevil of the University of Southampton, UK, being a leader in the field. Some of the injured found relief when incorporating copper into their diet or lifestyle.

Some of the treatments people experimented with were considered to be highly controversial, with two most notable ones being urine therapy and bloodletting, both of which had been used in traditional medicine before the pharmaceutical industry took over healthcare and are still used in other countries and cultures in present day. The more acceptably named "therapeutic phlebotomy" is used in modern American medicine for the treatment of specific blood disorders. Both urine therapy and therapeutic phlebotomy are the subjects of numerous scientific papers, with the latter being the subject of a study published in spring 2022 after it had been discovered that Australian firefighters had significantly decreased levels of toxins in their blood following donation.

Some vaccine-injured who had noted a significant improvement in symptoms immediately after having blood drawn, were keen to book more blood tests either through their doctor or privately purely for the purpose of getting blood out, and failing that, they explored traditional blood removal through ancient Chinese bleeding, wet-cupping, or leeches. Some even learned how to draw their own blood.

Bri tried to keep an open mind, and in some ways was inspired by the commitment to healing she witnessed. But she was also shocked at the extent to which vaccine-injured all over the world were experimenting with anything and everything that might help them. *This is what happens when you can't access good medicine—intelligent people are so desperate, that they are willing to stick needles in their own veins or drink their own urine. And these measures seem to be helping them.*

Some traveled to other countries to access treatments unavailable or unaffordable in their own: stem cell transplants in Mexico, or blood washing in Cyprus. Unbeknownst to React19, some overseas injured even came to America, in the hope of accessing treatments unavailable in their own countries, but without using React19's providers, had ended up deteriorating through frustration and overexertion, as well as lack of treatment.

Some people deteriorated after trying that newly discovered therapy that seemed to be helping their fellow injured. After putting what was left of their hope—and perhaps what was left of their savings—into a final treatment to try, only to then become even worse, life became intolerable for some. People who had perhaps been struggling with chronic illness for much of their lives before the vaccine were often vulnerable to this kind of situation. They had chosen to get the vaccine because they'd been told that it would protect them from deteriorating, only to become so much worse than they'd ever been before. They didn't have any more fight left in them, and instead chose to end their lives.

Finding treatments for the mental health impact of being vaccine-injured was an ongoing challenge. This wasn't a situation where sitting in front of a therapist, talking about a terrible childhood or traumatic event, then blaming your parents, was going to help in any way whatsoever. If anything, the chemicals that the body released while reliving historical difficulties made many injured feel even worse—it was as if the body had lost its ability to process those chemicals and they remained stuck inside the body, exacerbating neurological and cardiological symptoms. No amount of talking therapy would fix a fundamental problem with the body's filtration or immune systems.

But many of the injured did have medical PTSD, brought on by the way they might have been treated in a medical environment and also by what for some was a sudden, violent, and terrifying reaction to the vaccine. The experience of being taken away in an ambulance is distressing for anyone, and for those injured during 2021, they were usually admitted unaccompanied because of Covid restrictions. The environments were chaotic, and the fear of Covid ever-present. Depending on the level of kindness and compassion that was shown to anyone whose illness had occurred soon after vaccination, medical PTSD and health anxiety were understandable components of being vaccine-injured, which were frequently triggered by mention of the vaccine or sight of a needle in the weeks, months, and even years to follow. And the vaccine-injured were trapped inside bodies that didn't seem able to process any of that.

Bri had found EMDR had released her from that feeling of being trapped within an ongoing medical nightmare that could unexpectedly overwhelm her at any point. But she also found that her mental health—previously not something to which she'd needed to pay much attention—now required *ongoing* attention. Many of the injured were finding the same.

Mental health was the one thing for which the injured could find treatment, and it didn't need to cost them a thing. It wouldn't necessarily "fix" anything, but it seemed to prevent symptoms from getting worse, and made the symptoms that remained, at least manageable.

Breathing exercises, grounding, and meditation became a daily practice for Bri, along with many others who changed their entire routines to accommodate ongoing health management. Awareness of the impact of the basics—food and sleep—on mental health was also paramount; Bri even felt the difference in her mood if she decided to taste just a little bit of her daughter's cupcake. There was no room for cheating.

Many of the injured seemed to not only be sensitive to foods, chemicals, and even their own emotions, but they were also sensitive to others' emotions. There seemed to be rather a lot of empaths within the injured community across the globe, especially with those who were leading support groups. Understandably so—empaths want to take away people's pain.

But they're not always very good at setting boundaries, let alone sticking to them.

This was another component of mental health management that Bri was slowly learning about, with Joel and Kyle being her inspiration, and Jessica her coach. Running React19 and the constant demands people seemed to make on her—even coming from other injured people at times—was an ongoing challenge for Bri. Kyle was very self-aware and always seemed to know when to take a step back from advocacy and put himself first, and Joel simply refused to accept anything that would negatively impact him.

Treating the mental health of the vaccine-injured required a complex approach that even the most experienced of mental health professionals struggled with. It required an acknowledgment of the very *physical* impact of an emotional response, as well as the recognition that emotions were greatly impacted by chemicals . . . chemicals in the gut, the blood, and the brain . . . that the body was struggling to physically process. It also required understanding that the presence of those unprocessed chemicals in the body may well be triggering memories and feelings associated with past distressing experiences, completely unrelated to the vaccine; distressing experiences that had not negatively influenced the injured person's life in any way for many, many years.

Bri noticed that the injured who seemed to be the most effective at managing their own health seemed to be the ones who were the most willing to keep an open mind toward *all* the treatments that might be out there, even if they were from modalities that had never resonated with them before. Pharmaceutical or natural, physical or spiritual, chemical or energetic—all approaches seemed to have some level of benefit for the vaccine-injured, and a combination seemed to have the greatest impact.

But it was the philosophy with which every injured person approached their own treatments that seemed to be the most powerful tool at hand. A lifetime of expecting the healthcare industry to fix whatever health problem appeared had led to generations of people who didn't know *how* to take control of their own health. Most people genuinely didn't have confidence in their body's natural ability to heal; they didn't know *how* to create a lifestyle that could maximize that natural ability. They only knew to look to the pharmaceutical industry for a "cure."

When no cure existed—for how could one when your *condition* didn't even exist?—then it was up to the patients to find whatever treatments they could, and use them to create whatever quality of life they could. It was up to the patients to work out how to exist alongside their new health limitations rather than fight against them.

And that level of *active* acceptance wouldn't necessarily bring recovery, but it could bring healing.

CHAPTER 15

FINDING HEALING

The Covid vaccination rollout left a lot of damage in its wake. It wasn't just the vaccine-injured who were in need of healing. America's treatment of and response to Covid vaccine injuries were indicative of a country that was broken; a country that had forgotten how to operate with kindness and compassion. It was an America that was disconnected from humanity.

The disconnection had been growing long before Covid.

America's disconnection had been fueled by its media and politics for decades. Politicians had focused on the policies with which they disagreed rather than the ones they shared, prompting angry private and public arguments among families and friends rather than respectful discussions. Election debates centered more on criticizing the opposition and stoking hatred for the other party, rather than candidates explaining their own actions and aspirations in a way that inspired. If the country's leaders couldn't speak of and to each other with respect, why should its citizens?

Such public interactions between politicians made for good entertainment, picked up by a media hungry for anything that would generate more traffic, more viewers, more responses, and therefore more advertising revenue. The more people argued, the more money could be made. Social media platforms were perfect for vehemently expressing opinions, and users could disconnect from anyone with whom they disagreed, at the touch of a button. You didn't even have to use a real name—you could even disconnect from yourself.

This disconnection had led to a culture where it was possible—and acceptable—to instantly dismiss anybody with an experience or an opinion that didn't match your own worldview, even if it was someone you had loved and admired for much of your life. There was no room for being unsure, for nuance, for questions—you had to pick a side and stay there. Decide on your beliefs and act as if they were the one and only truth. Being right mattered more than being kind.

Then Covid happened, and most people's connections with actual humans were off limits, restricted, or quite literally, masked. Connecting with others was bad for your health; it could kill you, or someone you loved. "Other people" were something to be afraid of. Citizens were told to stay at home and, with only the mainstream or social media commentaries for company, the disconnection—the dismissal, the derision, and the discard of others—grew deeper and deeper. Creating the exact opposite of "society."

America hadn't been as divided since the Civil War.

It was in the middle of all this division that the Covid vaccine was rolled out. Rolled out—and mandated—in the context of a culture distinctly lacking in any kind of compassion and respect for anybody's opinion except one's own. And without any understanding that those opinions might evolve over time or through experiences.

For those who had been injured in the past by other vaccines, the Covid vaccination rollout had been incredibly triggering. Not only did they have to deal with the pressures of having to take another vaccination, but they watched everything happen with full knowledge that the Covid vaccine would cause harm to some. They also knew that the Covid vaccine-injured would be abandoned just as they had been; left to fend for themselves while the rest of the world pretended they didn't exist.

There were people who knew they didn't want the Covid vaccine, or at the very least wanted more time to make their decision. They were shunned the world over—by employers, colleagues, friends, and family—while their segregation from society was celebrated. And when mandates were lifted, they were expected to return to life as normal and interact with the very same people who had refused them a seat at Thanksgiving.

Some had been subject to direct abuse because of their personal views on vaccination, masking, lockdowns, or Covid. People who refused to wear masks had been spit on and called murderers by people who *did* wear them—people who considered breathing on each other to be unacceptable but spitting on someone to be fine. This was the kind of abuse that would take years to heal. But healing is hard, messy, and complicated. Healing requires vulnerability.

Instead, some of the unvaccinated became just as judgmental and prejudiced as the people who had treated them in such a way. Some chose to ridicule people who wore masks, who got vaccinated, or who believed what the media or government were saying. It was an understandable response to the abuse they'd suffered, but it ultimately served to further disconnect them from their loved ones as well as their own pain.

Most people who chose *not* to get vaccinated did so out of love for humanity. Yet it felt like humanity turned against them.

Most people who chose *to* get vaccinated also did so out of love for humanity. Many were pressured or outright bullied, but they ultimately made their decision out of love for their jobs, or their patients, or the families for which they needed to provide. Love could be a powerful motivating factor for many behaviors and decisions among the general public, even if it wasn't for the people in charge. But somewhere along the line, in the middle of all the fear and the pain and the isolation, love was forgotten.

Even the people who pushed the needles into arms did so out of love for humanity. The doctors and nurses who had been trained to promote and administer vaccines believed that they were playing their part in saving lives. So when they too, were subject to abuse and accusations of being liars and murderers, they emerged from the rollout with their own trauma to heal. They had worked all the hours they could under extraordinary circumstances, neglected their own families and risked their own health, only to be spat on too.

Some healthcare professionals would emerge from the rollout no longer sure about their own feelings about vaccines. Some had serious and disconcerting questions about the role they may have played in them throughout their entire careers, especially when they learned how the Covid vaccine-injured were treated. But who could they talk to? Talking about vaccine damage felt like a terrible betrayal of the profession to which they had dedicated their lives, yet *not* talking about it felt like betraying the people that profession was supposed to protect.

Healing from any trauma involves trying to make sense of one's own personal experience, predicament, and pain. Healing from the collective trauma of the Covid years involved trying to make sense of *others'* personal experiences, predicaments, and pain, something that had been actively discouraged for many years prior.

For some, being vaccine-injured opened the doorways to increased understanding and empathy in their lives; for some, they felt those doors slam in their faces.

Not everybody who had a vaccine-injured person in their life was kind to or about them; instead often oblivious to how much their words hurt, adding another layer of trauma from which the injured would need to heal. Repeatedly being told how many jabs a friend had taken, followed by the words "and I'm fine" would be like the Band-Aid on a wound being ripped off, over and over again, leaving the injured to wonder . . . *If I'd just told you about a miscarriage, would you tell me about all your healthy children?*

Watching close friends suddenly or gradually shut themselves off from your life at a time when you were feeling isolated enough was deeply hurtful and left the injured wondering if their friend's connection with being pro-vaccine perhaps meant more to them than their connection with the friendship. And the injured had to be honest with themselves—were they now dismissive of their friend's strongly held beliefs about vaccines without taking the time to understand what was behind them? Many a friendship was tested by the Covid vaccines.

Some people were able to put a vaccine injury—their own or that of someone they cared about—in the context of the entire history of their relationship, and thereby find a way to still lovingly connect despite the uncomfortable feelings that an injury raised. Love could provide comfort in the middle of that discomfort. Love was capable of being the reminder to value friendship higher than any pharmaceutical, government, or media messaging around Covid, the Covid vaccine, or vaccines in general. Or indeed anything. Friendship could be healing in itself—it could heal divides; it could nurture the connection that the modern world was destroying; perhaps it was *exactly* what was needed during conflict. Could people find the courage to move closer to each other during disagreements, rather than pull further apart?

And if people could come together in love over conflicting opinions on the Covid vaccine, could they come together in love over conflicting opinions about other things? The kind of things that had been driving people apart? Could actively seeking out connection during vaccine injury conflict inspire others struggling with similar friendship issues—perhaps relating to politics, religion, or sexuality? Could the vaccine-injured bring about a kind of healing that went beyond that of their own bodies?

New, or closer friendships would come from unexpected places and form part of the injured's healing. As friends and acquaintances learned about an adverse reaction, some were able to put aside their own feelings for or against the Covid vaccine and simply connect with being human. Some were able to focus unconditional love on a friend they could see was really struggling regardless of the reason why. Some took the time to pop by with simple, nourishing, fresh foods that they knew their friend could safely eat. Some emailed links of a person they'd come across on the internet talking about vaccine damage. Some sat quietly alongside an injured friend in the literal and metaphysical darkness—these were all healing acts of love.

Love had the power to transcend all divides.

But love would be put to the ultimate test for the vaccine-injured and their partners. Some of the injured would discover that not all of their relationships would survive, coming under the same pressures as other chronic illness sufferers do, and with the added taboo of being related to a vaccine. The ongoing symptoms with no cure in sight, along with the unexpected onslaught of medical bills that just one person's income was now responsible for, was enough to strain any relationship.

In some cases, the pressure was too much, and some of the injured would see their relationships end. Vaccine injury—just like any life-changing event—had a way of showing what people were made of. Some partners stepped up and stepped in, taking on their partners' responsibilities as well as continuing their own. Some stepped out, sometimes incredibly cruelly with no concern for how their newly disabled partner would manage, and adding to the levels of trauma and abandonment the injured person was already dealing with. When news would circulate within the community of an injured person's partner leaving them, that was when Bri would go on high alert—she'd seen too many suicides chosen by people who didn't have support at home. Partner abandonment would be the last straw for even the strongest among the injured.

Partners, parents, and close friends of the vaccine-injured had *no idea* just how much their support meant, often literally meaning the difference between life and death. They had the power to be part of the solution by creating more healing, or be part of the problem by causing more harm.

Those who loved the vaccine-injured were in need of their own healing. They still are. Countless people live with the memory of their loved one being taken away in an ambulance—often multiple times—and not being able to accompany them. They experience the daily trauma of watching their loved one's suffering, and they have had their own personal and professional lives completely changed by a vaccine injury. Yet they somehow find a way to exist within the ongoing reality of who their loved one now is, and what made them that way. And they do it all in the background, behind the injured.

The carers of the vaccine-injured are even more invisible than the injured themselves.

And then there are the vaccine-bereaved, for whom healing is an even more complicated task—attempts at which are made while fighting to get their loved ones' deaths recognized. Despite knowing that nothing they can do will *ever* bring back the touch of a hand, the sight of a smile, or the sound of a laugh . . . they fight on, carrying their heavy burden.

Ernest Ramirez somehow found the grief of losing his teenage son just that little bit easier to carry after meeting his new injured friends in Washington. He unexpectedly found strength in adopting the role of protector—as well as entertainer—of the injured, as he both gained comfort from and offered it to what he came to call his "injured family." His joking hid the pain that came flooding back, clearly visible, whenever he stepped forward to speak about his son. Despite the pain, many in the injured community would take joy from outrageous and sometimes macabre ways of making each other laugh, often with the kind of jokes that only another injured would understand.

The injured found ways to connect that were full of love and joy despite all the divisions that the Covid vaccine and everything that came before it had created. Despite all of their individual differences. And there were many. The injured were Republicans and Democrats and all those in between and outside of either. They were male and female and whatever else anybody might choose to consider was between the two. They were young and old, rich and poor, ethnically diverse, formerly healthy and formerly not, and representative of all cultures, religions, lifestyles, and any manner of beliefs, experiences, and professions. They had a multitude of topics upon which they could disagree, divide, and disconnect.

Instead, somehow they were managing to build connections among each other that transcended political or religious beliefs. Knowing first-hand how it felt to be abandoned, ridiculed, or disbelieved, many of the injured found themselves making conscious efforts so that their own actions didn't make others feel as discarded as they had. Online support group gatherings were full of people perfecting the art of simply *listening* to another person's pain; not trying to fix it, not trying to distract from it, not trying to argue with it, but instead creating space for it with the quiet strength and compassion they were trying to find as they worked on facing their own.

Could the Covid vaccine-injured, in the middle of the fragmented and fractious remnants of a once free and proud society, possibly heal divides?

Could healing come from efforts made to look at the world with love rather than hate? Could healing come with the assumption that most normal, everyday people really do just want the world to be a better place? Most people, deep down, want the same thing but somehow everybody had been so busy fighting about how to achieve it, and become so distracted by the fighting that they forgot what they wanted in the first place. Had Americans become so distracted by everything else that they didn't even know about the vaccine-injured?

Could the simple act of being able to talk openly about the vaccine-injured and all the complex issues surrounding how America treats them bring a step toward healing the country? Is that what America needed to do to heal its divides? Unite in the defense of the First Amendment so that the censorship of the vaccine-injured—or any other sector of society—could never happen again. To come together in the protection of fellow Americans' rights to share their lived experiences, to seek support for their illnesses, and to question government decisions. To be proud of the freedom that Americans had fought for to speak openly about their lives and their beliefs. To be able to turn to each other with a smile and say that they disagreed on the subject at hand, but together would celebrate living in a country where disagreement and dissent was allowed. Even if that disagreement and dissent were with presumed experts. Perhaps *especially* if that was with experts.

For some of the injured, their healing would remain very much tied up with what the experts—their doctors, scientists, or government—said about vaccine injuries. They aimed for a complete physical recovery; a return to life as it used to be; a return to being able to move, eat, drink, think, work, and sleep, just like they had before getting vaccinated. For some it meant aiming for a life where their first activity in the morning *wasn't* to check in with their bodies; where they didn't have to think about the repercussions of accepting an invitation; where family activities didn't have to be planned around treatments; and where vaccination wasn't a topic for which they had to choose their words carefully. For some, healing meant getting their old life back. Just the way it was. And almost four years on, with all her connections around the world, Bri didn't know a single injured for whom that had happened . . . not even close. Despite all efforts, some of the injured would struggle to find even the slightest bit of relief from their symptoms and situation.

For others, healing meant something different. Healing wasn't necessarily "recovery." Healing meant creating a new life. Finding ways to embrace those moments in the morning where they not only checked in with their bodies but also their minds and their souls; learning to look back on the adventures once had with a smile of appreciation, rather than the sorrow of loss; learning to sit *with* the pain in their minds and spirits as well as their bodies, rather than fighting it; and learning how to deal with unkindness with dignity and grace. Healing meant learning to be patient for—and perhaps even curious about—what might emerge from the darkness.

For many of the injured, healing had to come from within.

For those who made healing their own responsibility, healing would not come from anything a doctor could prescribe, a researcher could discover, or a government could finance. Those were all tools with the potential to bring about relief, recognition, or recompense, but the healing journey following a Covid vaccine injury was far more complicated than that. It involved facing up to everything you thought you believed in; facing up to everything that had been central to your worldview; facing up to everything and everyone you had trusted; and accepting that all was not as it had seemed. The institutions and individuals in which you had had faith no longer existed. Perhaps they never had.

Healing came from the breaking of the connection to the world as it had appeared *before* becoming vaccine-injured. It came from the disconnection from distractions that served no purpose to someone focusing on rebuilding their life. Or someone who was preparing for their own death.

For some, healing involved looking at death itself, perhaps exploring a new religious belief, perhaps rejecting an old one. It involved not thinking about the future—instead an acknowledgment that you might not necessarily have much of one, so taking life one day at a time. And finding beauty in that. It involved overcoming the fear of death, and for some, ultimately concluding that how one *lived* was more important than how one died.

Healing then came from the creation of new connections. Connections with nature and all it had to offer; connections with beautiful people, places, or produce. Connections with community, with kindness, with compassion. Connections with oneself. Healing grew from what was experienced by the mind, body, and spirit; and it grew from the connection between all three.

The healing of the vaccine-injured meant connecting with everything that it meant to be human. To be imperfect. Evolving. Learning. Making mistakes. And owning them.

For Bri, healing meant being honest with herself about the person she had been, the beliefs she had had, the life she had led before her injury, and the mistakes she had made. It meant reflecting on and facing up to the times when she might not have been as compassionate as she'd liked, especially with regard to vaccines. It meant sitting with the unsympathetic online response she'd offered years ago to the neighbor whose teenage nephew had been injured by the HPV vaccine. A response that had not made Bri especially proud of herself after the teenager had removed his own breathing tube in 2018. He had been a bright, joyous, smiling, loving young man who felt that his life was no longer worth living. And Bri—who considered herself to be very sensitive—had had no idea just how much her words at the time would have hurt. Now she knew.

The memory of the interaction *haunted* Bri during her own injury, and she often wondered how she could possibly make up for her thoughtless words. Then one day, having come across an interview with Bri, the aunt had reached out to her. Despite the way Bri had belittled her nephew's suffering in defense of vaccination, the aunt was gracious enough to say that she was glad Bri hadn't chosen the same path as her nephew and thanked her for still being here. It was this level of compassion that Bri wanted to carry with her into the rest of whatever life she had left.

Healing for Bri also meant recognizing her lifelong need to make everything better for everyone—her need to "fix" everything—and how much that trait could impact her health unless she learned to make herself a priority. It meant learning to accept that her mere existence brought the people around her joy—she didn't have to be rushing around organizing and helping everyone for people to want her presence in their lives. She could just *be*. And that would be enough. She was loved anyway.

This desire to help others was what had led Bri into advocacy, but steps had to be taken to prevent that advocacy from holding back her own healing. Leaders of other support groups around the world—many of them also fixers and empaths and healers at heart—would face the exact same challenge.

Bri's healing came in learning how to set boundaries, even if it meant saying "no" to other injured, many of whom viewed Bri with the utmost respect but sometimes forgot that she was injured herself. It came in learning to accept the fact that not everybody was going to like that—or her—and some would even actively dislike her. Some would be out to cause trouble for her, even those she had initially thought were allies. She would have to handle jealousy, rumors, and threats sometimes from the very people she was trying to advocate for, and this would hurt her deeply. She had to find healing from that too.

For Bri, healing meant connecting with both the pain and the joy that life brought. Being able to *feel* everything now, good and bad. In the absence of being able to find happiness in all the physical activities she used to enjoy—climbing, wakeboarding, skiing—she discovered that joy could be found everywhere, if she was open to it.

Bri found joy in connecting with nature, even if it was only from her own front yard. She could feel the warmth in the air that heralded an incoming storm, and she'd smile at the squeals of delight from Cooper and Hannah when they heard the thunder in the distance. The kids would come running into the house, knowing the drill. But instead of staying inside with them to wait for the hail to pass, Bri found joy in stepping out onto the grass, feeling the wet ground squish between her toes, and tilting her head back as ice landed on her face, her tongue, and her hands.

Healing for Bri meant embracing all those magical moments without giving too much space to regret, resentment, or rage, despite whatever physical and emotional pain she might be in.

Bri's healing came from the kindness within the friendships she made among the injured community—the injured, the bereaved, and the genuine allies. It came from the care of a pop star who was keen to protect Bri from the same kind of toxicity she herself had been exposed to; it came from Ernest joking about Bri being his bossy little sister. Healing came from the warm hug of a senator who refused to turn his back on her; and the sincere text from a presidential candidate reassuring her that he wouldn't forget the vaccine-injured.

Bri's healing would be halted every time there was another suicide; every one of them opened up the pain of the one before it. Every one took her back to losing Heidi.

Every suicide reminded her of how close she had been to choosing that path, something for which she hadn't quite forgiven herself—maybe she never would. She didn't judge anyone for choosing that path, but she often struggled with her own guilt associated with those dark days.

Bri tried not to judge anyone as she adjusted to her new life. She too, had had conviction in her beliefs prior to injury but had subsequently had so many of those beliefs proved to be misguided time and time again; they had been strongly held beliefs in politics, healthcare, media, and the world in general. All torn apart. All subject to dramatic change for anyone in the blink of an eye . . . or the touch of a needle. Bri learned to be comfortable with not needing to be "right" about any of those subjects anymore; they all seemed to cause more conflict than connection anyway. All that really mattered was love.

Despite how unlovable Bri had felt when she'd been unable to participate in life as it used to be, she had always felt *surrounded* by love. Brian had spent night after night simply holding his hand on her heart, carrying her up and down stairs, and reading countless scientific papers just to find any clue to what might help his wife heal. He had relentlessly pursued medical professionals all over the world, risked his career by speaking out in public, and carried the weight of everything that went on behind Bri's advocacy work. Bri was incredibly grateful to have Brian, and wanted so much to take away the pain and pressure that she knew he sometimes struggled with under their challenging circumstances. She longed for the days when the two of them could just have an adventure together again— climbing and skiing and exploring the beautiful outdoors that had connected them in the first place—and was starting to accept that her injury had led them on a different kind of adventure. At least they were on that adventure together.

She was grateful to have her mom alongside her on the adventure too. Marianne's unwavering support of her daughter during her time of need had somehow healed unresolved pain Bri had always had at the back of her mind, when reflecting on the challenges the entire family dealt with during the divorce. Bri took incredible comfort—and more than a little amusement—in the knowledge that her mom was sitting outside the Senate building, quietly focused on her knitting, while Bri was inside, trying to change the world.

Bri's relationship with Trish had in some ways gone on its own healing journey. During the early months, Trish had been very comfortable with Bri talking about her symptoms, but not the cause. As Bri had become increasingly public, Trish had become increasingly uncomfortable, to the point when the sisters stopped talking about the activism work it had led to. But they never stopped talking. They just found other things to talk about. Other things they cared about. At no point did Trish ever withhold her love for Bri, and at no point did Bri ever feel unloved by the sister she so admired. Their love for each other was more powerful than their personal beliefs and individual experiences.

Bri's personal challenge as a vaccine-injured person and advocate was essentially a challenge rooted in love. Love for her family, her community, and her country. It wasn't just about her own health, and it wasn't just about getting the other injured the help that they needed. It wasn't just about working out who were allies and who were not. It wasn't just about holding institutions and individuals accountable for the damage they cause. It wasn't just about finding the money to keep herself or anybody else going. It was about inspiring love. It was about having felt hate from all sides, and feeling layer upon layer of the pain of others added on to her own, yet ultimately deciding that the only way to see through those layers was through a lens of compassion. To commit to finding common ground despite any differences, and to work toward a world where people felt safe and secure enough to disagree respectfully, curiously, and lovingly. It was about creating the world that she wanted to live in. And that world was full of love.

If there was a way to share her story, and the stories of the vaccine-injured, that could somehow inspire more connection, compassion, and *love* in the world, then Bri would try to find that way.

It was worth a shot.

EPILOGUE

On May 13, 2024, Brianne Dressen filed a lawsuit in the District Court of Utah against AstraZeneca, for breach of contract in its failure to provide medical care. It was the first lawsuit of its kind in the United States. Brianne also joined Kristi Dobbs, Nikki Holland, and Ernest Ramirez to file a federal lawsuit against the Biden administration, US surgeon general, HHS, and the CDC, demanding an end to social media censorship of vaccine-injured Americans. *Dressen v. Flaherty* and *Dressen v. AstraZeneca* will begin deliberations in the Fall of 2024. Bri continues to serve as co-chair of React19, and maintaining her health is still a daily battle.

Joel serves as co-chair of React19 and continues to further its mission to help the Covid vaccine-injured physically, financially, and emotionally. He remains disabled due to transverse myelitis and autonomic dysfunction. He has filed a federal lawsuit to contest PREP Act protections. His federal CICP claim has been denied.

Kristi continues to balance work life and home life with her remaining issues from vaccine injury. Despite a measured restoration of her previous life, she remains dedicated to the cause and serves as React19's secretary.

Danice continues to advocate for others through support groups and working with researchers. She also leads the research team at React19. She continues to struggle with her vaccine injury.

Ernest travels the nation to raise awareness and accountability. He serves as community outreach officer for React19, and is participating in two federal lawsuits. He has left his son's bedroom untouched, including his son's Bible, which is open at the place it was on the day he passed away.

Kyle dedicated 2024 to compensation reform by serving as React19's lobbyist in Washington, DC. He continues to help others unlock their own approach to healing. After his injury ended his professional mountain bike racing career, he is now exploring future career pathways. His federal CICP claim was denied.

Jessica's singing and dancing career remains on hold due to her injury. She is focusing on being a mother and finding pathways to restore her health. She has dedicated her future to advocating for others in her situation and continues to leverage her network in entertainment to shine a light on this issue.

Candace has worked tirelessly to restore basic life skills following her injury. Over three years post-injury, she was finally able to drive again. She continues to work full-time as a senior vice president of technology in a large company, while running her own hypnotherapy and life coaching practice. Candace offers a range of healing and personal growth programs designed to support others in the injured community in their journeys toward recovery and transformation. Brianne is a participant in her programs.

After her injury, Sheryl accomplished her dream of working as a therapist. However, after many months of struggling with a typical caseload, her doctors reduced her work to part-time. Her post-vaccine neuropathy and new autoimmune conditions continue to progress as she searches for answers from the broader medical community. She is still there to laugh or cry anytime with Bri.

Maddie has not regained muscle control even after intense therapy for up to six hours at a time, five days a week. She still has problems with her bladder and bowel functions, sweating, and POTS, among other conditions. She can now swallow small amounts of food but gets very sick afterwards. Steph is focused on sourcing physical rehabilitation and other therapeutics that will restore her daughter's health and quality of life. She continues to be an advocate for Maddie's healing and is determined to find answers. Nearly four years after her child was injured in the clinical trial, Maddie's mom has yet to hear from the CDC, FDA, or Pfizer. Her federal CICP was denied. Maddie has remained positive despite what she has been put through and is determined to recover and walk again.

In a letter dated August 26th, 2024, to the House of Representatives, Facebook's CEO Mark Zuckerberg stated that the Biden administration had repeatedly pressured the social media platform into censoring certain content related to Covid. Zuckerberg expressed regret at his company's policy changes in response to those pressures and admitted that they had made some decisions that they wouldn't make again.

Pascal Soriot, the CEO of AstraZeneca, earned $21.3 million in 2023, making him the highest paid CEO of the major European pharmaceutical companies. He received a knighthood from Queen Elizabeth II in 2022.

A LETTER FROM BRIANNE

My beautiful babies,

Our first moments were clear, I never knew this kind of love existed. As your mother my purpose is to love, guide, and protect you.

I want you to know that I did this for you, for our family. To teach you that through love, we can conquer anything. That we have a duty to help our family and community through hard times. I did this to protect you.

What happened to me was unexpected to say the least. My fight to live each and every day, I know you see it. I know you feel it. It is something I feel with every bone in my body, with every breath: in my attempt to protect you, I have instead opened your innocent lives to a Hell none of us ever knew existed. In doing so I have failed to protect you.

I hope that one day you may see that your mom didn't just sit on the couch all day. I hope you can understand that each and every day I was fighting fiercely to try to protect you still. To protect you from being exposed to the pain driving my mind to want nothing more than to scream out in agony for hours on end, and instead my attempt to mask this with a smile.

The worst punishment of all, is to see just how this hurts you. I see how deeply this has cut your hearts that were trained to know nothing but love and trust. I feel it with you.

I hope you can see the effort it took to sit beside you, as my body surged with electrical sensations, to stay present and brush your beautiful

223

hair from your face as you told me about your day. I hope you can see someday that only a mother's love can push someone through this level of discomfort, to escape the nightmare if even for a fleeting moment of joy with their child. In hopes that you can have this moment, but I know you know I am hurting. You tell me often you want me to be better, it's as if you will never ever be able to really exhale again until that happens.

I hope you can someday see beyond the resentment that will come with your teenage years, the repeated disappointments of all the things I couldn't do with you, and remember that with all my might I fought and clawed to get there, I did everything I could to be there for you.

I hope you now can see that I am still here, still fighting the exact same battle that has trapped me almost four years ago. The battle today, is to protect you. To protect you with every fiber of my being, from the world that you soon will also face. To protect you from those who hurt your mommy, to ensure they never ever can hurt you too. I hope you see now what I see, that I now see what protection you truly need. That which I would not have seen before. With all I have left, I will stand between you and that power that grips the world to ensure they can never ever hurt you beyond what they already have. I will continue to fight each and every day to make sure you, my beautiful babies, are loved, guided, and protected.

Love, Mom
Summer 2024

ABOUT THE AUTHOR

Caroline Pover is a multiple award-winning author, entrepreneur, public speaker, and philanthropist. Having spent a significant part of her adult life in Tokyo, her earlier works related to Japan. Her book about the decade following the Japanese tsunami and the forgotten community she befriended and supported there was awarded Best Memoir in the 2021 Next Generation Indie Book Awards. After being diagnosed with her own adverse reaction to a Covid vaccine, her writing became more health-focused. She produced the first survival guide for Covid vaccine injury, followed by a health workbook. *Worth a Shot?* is her eighth book.

Caroline is the Chair of Trustees of UKCVFamily, the only registered charity in England and Wales supporting the Covid vaccine-injured. She lives in the Cotswolds, UK, where she runs a food business in between writing, public speaking engagements, and snuggling her rescue dogs. Her website and blog can be found at www.carolinepover.com.

Some Other Books by the Author

Covid Vaccine Adverse Reaction Survival Guide: Take Control of Your Recovery and Maximise Healing Potential

One Hundred Days of Healing: A Workbook for Sickness, Separation, and Sorrow

One Month in Tohoku: An Englishwoman's Memoir on Life after the Japanese Tsunami (Winner of Best Memoir, The Next Generation Indie Book Awards)

Being A Broad in Japan: Everything a Western Woman Needs to Survive and Thrive

WHAT CAN YOU DO?

Donate to React19, UKCVFamily, or any other charity supporting the Covid vaccine-injured.
Purchase copies of this book to give as gifts.
Ask your local bookstore and/or library to stock this book.
Start compassionate discussions around you.
Suggest this book to your book club.
Teach this book in your classes.
Invite Brianne and Caroline on your podcasts.
Write to your elected representatives.
Write about this book on your social media pages.
Review this book on Amazon or any other book review platform.
Actively create the world you want to live in—make it a world full of love.

Get Informed
Visit These Sites and Get This Book
React19: https://react19.org/
UKCVFamily: https://www.ukcvfamily.org/
Covid Vaccine Adverse Reaction Survival Guide: Take Control of Your Recovery and Maximise Healing Potential by Caroline Pover